FOOD SANITATION

THIRD EDITION

Rufus K. Guthrie

Professor of Microbiology and Ecology
The University of Texas
Health Science Center at Houston
School of Public Health

An **aVi** Book
Published by Van Nostrand Reinhold
New York

An AVI Book
(AVI is an imprint of Van Nostrand Reinhold)
Copyright © 1988 Van Nostrand Reinhold

Library of Congress Catalog Card Number 88-10741

ISBN 0-442-20544-9

Printed in the United States of America

Van Nostrand Reinhold
115 Fifth Avenue
New York, New York 10003

Van Nostrand Reinhold International Company Limited
11 New Fetter Lane
London EC4P 4EE, England

Van Nostrand Reinhold
480 La Trobe Street
Melbourne, Victoria 3000, Australia

Macmillan of Canada
Division of Canada Publishing Corporation
164 Commander Boulevard
Agincourt, Ontario M1S 3C7, Canada

16 15 14 13 12 11 10 9 8 7 6 5 4 3 2 1

Library of Congress Cataloging in Publication Data

Guthrie, Rufus K.
    Food sanitation / Rufus K. Guthrie.—3rd ed.
       p. cm.
    Bibliography: p.
    ISBN 0-442-20544-9
    1. Food handling. I. Title.
    TX537.G85 1988
    641.3'028'9—dc19   88-10741

Contents

Preface to the Third Edition

The original stimulus for writing this book came from requests of students in Sanitation Short Courses for Cafeteria Workers at North Texas University, Denton, Texas in the summers of 1968 and 1969. I am grateful to them for pointing out the need for such a book. Because of that initial impetus, the work has been and still is focused primarily on the foodservice industry, and it is hoped that this edition will be even more useful to workers in this field.

Since publication of the first edition in 1972, the science of sanitation has changed very little. However, many changes have occurred in food science, eating habits, and dietary practices since that time. Several discoveries, or at least changes in emphasis, have been made in the field of microbiology that have a direct bearing on food sanitation. The changes that have occurred make it increasingly important for the homemaker, as well as the industry worker in food production, processing, and service, to have a good understanding of the principles and practice of food sanitation and an awareness of the specific dangers to human health that result from the failure to practice good sanitation. As the number of meals consumed away from home increases throughout the world, sanitation at all stages of food production, preparation, and service becomes more critically important. This edition attempts to provide a basis of factual knowledge that can be imparted easily to the food industry worker. Although turnover in this industry is particularly high, any worker equipped with this knowledge need not endanger the consumers of food in any environment.

In this edition, materials have been updated that deal with food and water regulations, foodborne disease outbreaks, microorganisms that cause foodborne diseases, food additives, chemical contaminants in food, good manufacturing practices, retail food store sanitation, and training of food service managers in sanitation. Tables have been added to describe and identify disease-causing microorganisms, and updated municipal codes and laws have been included in the appendices.

It is a pleasure to again express my appreciation for my professor, the late Dr. Kenneth L. Burdon, whose example encouraged me in my microbiology career and inspired me toward writing this book.

I also wish to thank those who have given generously of their time and effort in secretarial help and patience in the typing, proofing, and editing of manuscripts of all three editions of this book, with particular thanks to Martha Pruitt at Clemson University, and to Peggy Donnellan at The University of Texas School of Public Health; and to express my appreciation to the contributors to the first two editions of this book: Dr. F. D. Borsenik, Mr. G. V. Brauner, Dr. E. M. Davis, and Dr. W. J. Harper for their part in making those editions successful. The assistance and encouragement of Dr. Donald K. Tressler in the preparation of the first edition of this book is gratefully acknowledged.

This third edition is dedicated to my wife, Maxeen, and my children, Annie and Kent, who provided the stimulus for my career in their love and patience with me, and to the students that I have had the opportunity to teach, whose intellectual curiosity has made teaching so rewarding.

Sanitary Science

The science of sanitation does not change rapidly; however, the emphasis placed on certain aspects of the practice of that science changes periodically. Regardless of emphasis and of the practical aspects of the practice of sanitation, the importance of preventive medicine and of sanitary practice continues as a universal concern. Medical care costs increase more rapidly than other living costs, and the attitude that it is easier to treat disease than to prevent it cannot be suffered. As a matter of fact, in many parts of the world, it is almost impossible to diagnose much of the disease that does occur, and treatment is not available in those areas once the disease is diagnosed. In developed as well as developing countries, improving and maintaining human health is vital, and treatment remains a back-up system to assist in restoring health when there is a breakdown in prevention.

There is almost universal concern with the safety of our food and the safety of the processes used to prepare and provide that food. We are frequently told that some constituent in our food, or the kind of food, or the amount of that food we consume, is dangerous to our health. We are bombarded with articles and books concerned with food additives, food sources, the role of diet in heart disease, cancer, children's hyperactivity, and so on. We are told that vitamins are essential to our well-being and that an excess of these same vitamins can be dangerous to our health. This concern with the safety of our foods has paralleled the dramatic changes that have occurred in the past twenty to thirty years in the types of foods we eat and food-preparation methods. In the developed countries, most of the food now consumed has been subjected to some degree of processing before it reaches the consumer. And in recent years, we have come to consume a greater percentage of our food away from home, thereby requiring consumption of even more food that has been processed and prepared by others. These changes in food habits have increased our concerns for food safety. We have reason to believe, however, that they have also increased the incidence of diseases caused by consumption of unsafe foods.

Civilization continues to be more aware of the numerous environmental factors that may adversely affect human health and well-being, and in no case is this more evident than in the production and provision of food to populations in all parts of the world. Nevertheless, a large proportion of adverse environmental factors have resulted directly or indirectly from human activities, some of which were originally intended to *maintain* the public health. An example of this paradox is one of the most successful public health practices ever introduced: the chlorination of drinking water. When chlorination, together with filtration, of water was begun, the incidence of waterborne infectious disease was dramatically reduced. Typhoid fever, once a dread and major killer in the United States, has become relatively rare, and routine immunization for its prevention is no longer deemed necessary. One can now consume treated water in the United States with little fear of contracting waterborne infectious disease. In recent years, however, we have come to realize that chlorine may combine with certain organic compounds found in water and form harmful, perhaps even carcinogenic, compounds. In recent years, we also have realized that the installation of water-treatment facilities and the continued use of these facilities requires more than routine or minimal attention. As a result of complacency following routine treatment of public water supplies, we have actually had an *increase* in the number of outbreaks of waterborne infectious diseases in this country, either as a result of breakdowns in treatment systems or negligence in the proper application of treatment to water supplies.

In view of these facts, we face a double dilemma. Should we continue to chlorinate our drinking water to prevent infectious disease and run the risk that a small number of individuals may develop cancer in the rather distant future? Or should we forgo the proven preventive benefits of chlorination in avoiding infectious diseases in large numbers of individuals to avoid the risk of contact with potential carcinogenic chemicals that may or may not be present in the water supply? Obviously, our choice should result in the greatest good for the largest number, but we do not stop there. Rather, we look for newer sanitary practices that will be as effective as chlorination in making our drinking water safe, and that will not create risks for even the few. We also look for ways to prevent the organic chemicals from entering the water supply or for ways of removing them so that they cannot react with chlorine to produce dangerous chemicals.

Humans have always attempted to manipulate the environment in order to produce the most advantageous conditions for mankind. An example of such manipulation is the practice of sanitation. *Sanitation* is a term that has traditionally pertained to health, particularly to human health. Human health, we have come to realize more and more, is affected by the total environment, including other living organisms, and therefore, in practice, sanitation includes maintaining the health of all living organisms. Broadly speaking, *sanitary science* is the application of those practices that will assist in improving, maintaining, or restoring the environment of the human so that it is most advantageous for maintaining good health. In the practice of sanitary science, then, those factors in the environment that may be directly or indirectly detrimental to our well-being must be manipulated.

Properly practiced, sanitary science is one example of good environmental control. With the objective of adding to our well-being, sanitation has come to imply much more than cleanliness and harmlessness. Proper sanitation has come to include many facets of the daily human experience that often go unnoted. Such practice of sanitation results in convenience to the public and improvement of the aesthetic qualities of homes, schools, public buildings, parks, cities, and all our surroundings. Sanitary practice may be shown to assist in maintaining the biological environment, thus aiding in the efforts to decrease the effects of pollution and restore our ecology to a balanced state.

While the need for sanitation can by no means be limited to the sanitation of our food supply, food sanitation is a major concern of sanitary science, simply because so much of our environment at some time touches, directly or indirectly, our food supply. Recognition of the importance of sanitation to the food supply came early in human history, with attempts at food preservation through drying, salting, and other techniques. The fact that the sanitary condition of food affected human health may have led to some of the early bans on certain foods by some populations, such as the ban on pork by some religions (because it is recognized that consuming pork that has been improperly produced, handled, or cooked is a means of transferring infectious diseases to humans).

Sanitation of drugs and cosmetics is closely allied to sanitation as practiced in handling foods, because drugs and cosmetics contact the human body, either internally or externally, when used. Similarly, food sanitation is also related to environmental sanitation because the condition of the environment affects food as it is produced, handled, stored, prepared, and served. Sanitation, in any sense, cannot be separated from the sanitary practices and personal hygiene of the persons who must handle the food. Sanitary science includes food sanitation, personal hygiene, and general sanitation, anywhere the human lives, works, or plays.

FOOD SANITATION

At one time in human history, as people established homes and ceased to be nomadic, food sanitation was a concern only in the home. When increased social activities became a part of human existence, the importance of sanitation, and particularly food sanitation began to increase, because of the increased concentrations of people. Communal meals became a part of life, and this practice has continued to increase to the present. In early days of community gatherings that included meals, many large epidemics occurred because of the lack of understanding of the principles of sanitation within the social group. In more recent times, as meals furnished to the public by individuals became more common, the recognition of the need for standards to protect consumers became apparent in light of the numerous incidences of foodborne disease outbreaks (Table 1.1). Recognition of this need led to the

Table 1.1. Foodborne Disease Outbreaks Reported in 1975 and 1980

Location	1975	1980	Location	1975	1980
Alabama	1	3	Nebraska	3	7
Alaska	4	5	Nevada	4	0
Arizona	2	4	New Hampshire	2	3
Arkansas	2	3	New Jersey	12	10
California	41	39	New Mexico	1	14
Colorado	1	5	New York	128	175
Connecticut	9	22	North Carolina	0	7
Delaware	1	1	North Dakota	0	1
Dist. of Columbia	0	1	Ohio	0	11
Florida	30	14	Oklahoma	3	1
Georgia	17	5	Oregon	7	3
Hawaii	15	39	Pennsylvania	21	28
Idaho	0	1	Rhode Island	2	2
Illinois	12	8	South Carolina	9	4
Indiana	4	0	South Dakota	1	0
Iowa	1	2	Tennessee	3	9
Kansas	0	8	Texas	17	30
Kentucky	8	5	Utah	3	1
Louisiana	15	3	Vermont	0	1
Maine	0	0	Virginia	4	11
Maryland	2	11	Washington	44	57
Massachusetts	8	10	West Virginia	0	1
Michigan	5	9	Wisconsin	13	11
Minnesota	25	13	Wyoming	1	1
			Total outbreaks	494	596
Mississippi	1	1			
Missouri	8	2			
Montana	3	3	Total cases	18,260	13,791

Sources: CDC Foodborne and Waterborne Disease Outbreaks, 1975, 1980.

enactment of laws regulating the levels of sanitary requirements. This development did not occur until well into the current century, when we began to regulate and define accepted standards of sanitation. The United States Congress may be said to have begun the science of food sanitation in this country with the passage of the Food, Drug, and Cosmetic Act in 1938 (hereafter referred to as the Act). Continued changes in our ways of life since that time have produced the need for more exacting requirements and new practices leading to many amendments of the original act, as well as passage of numerous new laws that affect foods from production to consumption. This includes, in addition to the federal laws, the passage and continuous update of state and local rules and regulations dealing with foods.

Besides establishing standards, laws dealing with food, drug, and cosmetic sanitation must also establish the limits and definitions of sanitary products. This is usually done not by stating what sanitary products are, but rather, what sanitary products may not be, or may not contain. It must have been obvious to the original authors of the Act that there were many ways in which food might be or become detrimental to consumer health. The Act considers food sanitation in terms that appear to make *unsanitary* and *adulterated* almost synonymous. Such usage is in recognition of the fact that sanitation involved the prevention of the spread of infectious disease, and the chemical and physical characteristics of food that may affect human health. The Act defines food as adulterated if it

. . . consists in whole or in part of any filthy, putrid, or decomposed substance, or if it is otherwise unfit for food; or if it has been prepared, packed, or held under insanitary conditions whereby it may have become contaminated with filth or whereby it may have been rendered injurious to health; or if it is, in whole or in part, the product of a diseased animal or of an animal which has died otherwise than by slaughter.

The complex character of this statement may be recognized upon closer examination. Terms such as *filthy substance* and *filthy* may actually refer to the physical state of the food in that what we may consider to be filthy may, during processing, be rendered harmless to the health of the human; yet the presence of such substances in food would indicate that the food was at one time potentially harmful. The purpose of requiring that food not be exposed to "filth" is to help ensure that the food be safe for human consumption. In recent years, numerous reports, popular books, and items in the news media have focused on the issue of food safety. The most frequently discussed items dealing with foods in these reports are food additives, the role of diet in hyperactivity in children, the role of diet in causing heart disease and in

causing cancer, and the importance of vitamins (either one or two fa-
vorites, or multiple vitamins). The general public's preoccupation with
food safety has grown in recent years, along with the many changes
in the nature of our food supplies, and with changes in our eating habits.
In a relatively few years, population concentrations have shifted to
urban sites, and there has been a concentration of food production
within smaller and smaller areas of the world. In this country, less
than five percent of the population produces not only all the food consumed by our country, but a large amount of that consumed in other
countries as well. In addition to the changes in production, there have
also been dramatic changes in the systems for processing and distrib-
uting our food supplies. These changes have resulted in greater varieties
of foods in different forms, and have increased accessibility to these
food varieties. Over half of all the food consumed in this country is
processed in some way before it is distributed. That processing is responsible for the great increase in varieties of food items now available.
Perhaps the greatest change in recent years, however, is the increase
in consumption of prepared foods that are purchased and/or consumed
away from the home.

Thus, although filth in food is still of concern, with the regulation
by governmental agencies involved, this aspect of food safety is not
encountered as frequently as it was in the past. The chemicals present
in food represent a constantly changing picture in the technological
world of today. Although chemical content of foods is covered in terms
of adulteration if a food has been treated in such a manner "whereby
it may have been rendered injurious to health," more recently a number
of laws, regulations, and ordinances have come to spell out most explicitly what can and cannot legally be present in foods. Chemicals
present in foods are of four different types, depending upon where the
chemical came from and when it was added to the food. The chemical
may be among

(1)   those making up the food product that are present as a natural
        part of the food and not as the result of any human action;
(2)   those that have come in contact with and remained in or on the
        food because of some event in the production, processing, stor-
        age, packaging, or preparation;
(3)   chemicals intentionally added for the purpose of modifying ap-
        pearance, flavor, stability, texture, or preservation; or
(4)   chemicals produced within the food by reaction of some food
        component with either an accidental or intentional additive, or
        by the growth of microorganisms that have contaminated the
        food.

The first class of chemicals, in most foods of good quality, are desirable and, hopefully, nutritious. Modification of these would be neither needed nor desired. The second type, those accidentally or incidentally added to the food can be harmless, somewhat detrimental, or frankly toxic. If harmless, there may be no specific prohibition of this type of chemical, but in the interest of avoiding adulteration, it should be avoided. If even somewhat detrimental to the human consumer, the chemical is prohibited as an adulterant, because the food would "have been rendered injurious to health." If prohibition is not practicable for this type of chemical, then in some cases strict limits or concentration tolerances are set. The third type makes up by far the greatest group of chemicals that we would expect to deal with in good quality food (other than the natural constituents). Recently, public interest in food additives has increased because of the emphasis that many media reports have given to some of these chemicals. In addition, regulatory agencies at various levels of government, particularly the federal government, have frequently modified laws and regulations to control potentially harmful food additives, and to eliminate those that are dangerous. Even some additives that have been in general use for a number of years are being re-examined and the potential for danger is being reassessed. However, there are still only two broad categories of food additives: those that have been scientifically investigated and whose use is regulated by the Food and Drug Administration, and those that are generally regarded as safe (GRAS substances) by expert judgment because they have been in use for many years, and no detrimental effects have been observed during that use.

The fourth type of chemicals result from biological factors associated with the foods, and are included in the "filthy, putrid, or decomposed substance," "if it has been prepared, packed or held under insanitary conditions whereby it may have become contaminated," and "if it is, in whole or in part, the product of a diseased animal or of an animal which has died otherwise than by slaughter." Generally speaking, most people would consider any "filthy" substance to contain large numbers of microorganisms (bacteria and fungi, at least), as well as any number of other creatures and substances that the mind may imagine. The presence of large numbers of microorganisms, even if the food has been held at low temperatures, may well have presented an opportunity for the food to undergo some process of decomposition or putrefaction that could make it undesirable for consumption. Even limited amounts of decomposition or putrefaction of food materials means that there has been alteration of the chemical and physical state of the food, and it is certainly reasonable to assume that some harmful chemicals may have been produced by the processes.

The preparation, packing, or storage of foods under conditions whereby contamination may have occurred again involves biological as well as chemical factors. Contamination may mean that poisons such as insecticides, disinfectants, and rodenticides have come into contact with the foods. If these chemicals have contaminated the foods, then the foods may be harmful to human health. Strict concentration limits in foods are set forth in laws and regulations for any of these chemicals. Contamination may also mean that mice, rats, cats, dogs, flies, roaches, weevils, or other insects or animals have contacted the food and deposited filth, hair, dead insect parts, or large numbers of bacteria, fungi, or possibly viruses and animal parasites that can produce diseases in the human.

Until the recent past, many people scarcely trusted foods prepared in places other than their own or their friends' homes. In far too many cases there have been, and continue to be, too many unscrupulous persons offering partially decomposed foods for sale and consumption, or offering for sale diseased animals, possibly even those that have died of disease. Many plant foods are shipped after some degree of spoilage may have begun, although this may be unknown by the seller at the time of shipping. This inadvertant sale of spoiling food is usually easy to detect and to correct; however, in some cases the condition of foods may be disguised by the preparation methods, including generous application of spices to mask suspicious symptoms of spoilage. The methods of preparation may even be used to mask physical changes in the food so that texture cannot be used to detect spoilage. Rather sizeable particles of foreign matter may be found in bulk sales or shipments of foods.

In application of laws regulating the sanitary state of foods, the practices just mentioned are the kinds that are of great concern, although these practices are now much more easily detected by technical methods than they once were. Persons who practice such deceptions must be removed from any position in which they might continue such dangerous activities, and foods that are so treated must be detected, confiscated, and destroyed before reaching the consuming public.

In recent years, greater concern has been evidenced with the less obvious aspects of sanitation, particularly concerning microorganisms, and also those aspects that are concerned with chemical contamination of foods. Reasons for this redirection of interest are changes that have occurred in the processes of food production, processing, marketing, and preparation, and changes that have occurred in the technology of detecting both chemical and biological contamination. Practices in all of these areas have undergone more than one revolution in the past twenty to thirty years. There has been an unprecedented shift from the local production–home consumption cycle of foods to large-scale commercial production, processing, marketing, preparation, and consumption of foods, with increased efficiency in local control of standards and sanitation of all steps in the process. Perhaps the greatest of these developments has been the change from home consumption of foods to greater emphasis on mass feeding establishments such as restaurants, cafeterias, and the dining rooms of a large variety of institutions. With this evolution, the necessity for emphasis on standards of sanitary practices dealing with microorganisms or chemicals has increased.

The requirements for the elimination of "any filthy, putrid, or decomposed substance" are, in final effect, requirements to control the microbiological characteristics of foods. Filth includes microorganisms, many of which may produce infections in humans, and many of which may produce a putrid or decomposed state in various food products. A lack of sanitation resulting from contact of insects, rodents, or any other animal with food or food products is likely to end as a microbiological problem because the mere contact of the intruder with the food will serve to introduce microorganisms into the food, even though the presence of such animal contact is not obvious. The presence of animal filth may be grossly apparent, in which case there is sufficient basis for declaring the product "unfit for food." Even if the contact of animals with food is inapparent, microbiological methods are usually successful in detecting and identifying the filth that may have been deposited.

Because humans must handle and be closely associated with the production, processing, marketing, and preparation of foods, much of the filth introduced into the food may be of human origin and originate in incidental contact with the foods. Even in this case the successful detection and identification of the contamination is frequently dependent upon microbiological methods. It is therefore apparent that the field of sanitary food science deals largely, although certainly not entirely, with microbiology. Because many diseases are produced only by microorganisms or their products, the production of putrid or decomposed states by microbial actions on foods, and the rendering of materials unfit for consumption by microbial activities, application of the principles of microbiology to sanitary science becomes essential.

WATER SANITATION

The very existence of life on earth depends upon the availability of a sufficient supply of water, and fortunately for humans, water is present in huge amounts in many parts of the world. However, it is not found in adequate amounts in every vicinity of human habitation. Concern for sanitation of food supplies is futile if one is not also concerned with the adequacy and sanitation of water supplies.

Historically, water has been the carrier of more diseases to the susceptible human than has food. The limited locations and extensive use of water sources for the human has been in part responsible for disease transmission by water. Most of our water supply must come from surface sources that are subject to accidental contamination by anything living or dead, in the air above, and on the earth surrounding the water. This accidental contamination is compounded by the use of surface water as a means of removing waste materials, which are detrimental, or undesirable to humans, with the net result that most of the surface water supply on earth is unfit for human consumption without treatment to remove and control contamination that humans have caused. Obtaining safe, reliable supplies of water for human use and consumption is one of the oldest continuing problems facing mankind. Gradually, we have awakened to the fact that we must attack the problem of a suitable, sanitary supply of water in more than one way. We must regulate and reduce the use of surface waters for removal of waste materials, and we must treat the water to eliminate any contamination that presents a danger to human health. We have studied the problem for a number of years and can see that we have made some headway toward solution, but have come nowhere near the perfect answer.

In 1974, the U.S. Congress enacted Public Law 93-523, The Safe Drinking Water Act, which authorized the Environmental Protection Agency to establish federal standards for the protection of the human from all harmful contaminants. This act also established a joint federal–state system for assuring compliance with these standards and for protecting underground sources of drinking water. Amendments to the act in 1977 required revisions of previous studies by the National Academy of Sciences "reflecting new information which has become available since the (Academy's) most recent previous report" and the Amendments require that the Academy's studies shall be reported to Congress each two years. In the process of continuing those reports to Congress, the National Academy of Sciences has now published the fifth volume of *Drinking Water and Health* dealing with contamination of drinking water supplies and the necessary treatment of those supplies to remove that contamination.

Large-scale epidemics of waterborne infectious diseases have been largely reduced as a result of the treatment of water supplies for human consumption. Yet, in recent years, there have been periodic increases in waterborne disease outbreaks which serve to remind us that we must be constantly alert to the proper treatment of water supplies for consumption. Proper treatment of water supplies has traditionally consisted of three steps: flocculation (coagulation), filtration, and chemical treatment (chlorination), which are discussed in more detail in chapter 6.

These treatments render "sanitary," or potable water of a satisfactory standard when the water is treatable, and when the treatments are properly executed. Potable water no longer contains those components that could be injurious to health. It may still contain many dissolved chemicals, and even some relatively harmless bacteria, yet it is safe for human consumption and use in practically all instances. To assure this safety, cities, towns, and states routinely check these treated waters at regular and frequent intervals to ensure that no breakdown in treatment or in the distribution system has occurred that would allow recontamination of the water.

Water of the quality obtained through these treatments must be used in food establishments, which are regulated by governmental agencies. Many municipal ordinances require that water coming into a food processing or service establishment be obtained from an approved source. This, for all practical purposes, means water that has been properly treated and that is regularly tested to assure its quality.

Although treated waters are completely satisfactory for the general purposes of consumption, food preparation, and ordinary cleaning processes in the United States, some operations require additional chemical additives. In the initial cleaning of foods that are heavily contaminated with filth, and therefore with microorganisms, cleaning with consumable water alone is insufficient to make the food safe. An example may be seen with pecans. Such nuts, during harvest, are likely to be heavily contaminated with soil, or even with animal feces. These contaminants must be cleaned off before the shell is opened to remove the meat. Consumable water will remove the soil, but will leave numerous microorganisms in crevices on shell surfaces. To remove these huge numbers of bacteria, fungi, and parasites, a large concentration of chlorine must be added to the water. Because chlorine is readily absorbed by the organic matter on the shell's surface, it is ineffective in killing disease-producing microbial cells unless the concentration is sufficient to both react with the shell surface and also leave an excess to kill the microorganisms on that surface. Following this washing, the pecan shells are sufficiently clean that the meats will not be contaminated simply by breaking the shells.

Water sanitation requires that sufficient contamination be removed, and that sufficient chemicals be added both to make the water harmless to humans and to do the cleaning job required. The treatment required, and the amount of chemicals added will vary with the condition of the raw water, and with the intended use of the finished water. Fortunately for humans, the nature of water is such that although it may be heavily contaminated with impurities of every description, it can be made safe for human use and reuse so long as we are willing to pay the cost of treatments required.

HOME SANITATION

The general application of sanitary practices in the home is a necessary ingredient to safeguard individual human health. If good sanitary or hygienic practices are used here, it will do much to overcome deficiencies elsewhere, and will form the basis of personal hygienic practices that will carry over and benefit the individual in all activities, in or outside the home.

Sanitation in the home must begin with housing that is of adequate standard to be made sanitary. A major consideration in this case is adequate space for the number of individuals involved, and for the necessary activities expected in the home. A shortage of adequate space for individuals in human habitations is perhaps one of the most serious problems found in developing countries worldwide and remains a serious problem even in the United States. We may think of plumbing and an adequate and safe indoor water supply as essentials, but homes can be maintained as sanitarily adequate, and have been so in the past, without these conveniences. Most assuredly adequate plumbing and a safe indoor water supply do make it much easier to maintain sanitation in the home, and in the majority of homes in the United States these are accepted as basic necessities.

With adequate space and the basics of convenience assumed, the consideration of home sanitation becomes a matter of the basic principles of sanitation, which are the removal of filth from the environment, and the prevention of human contact with filth. First and foremost, then, is the requirement for personal cleanliness of individuals. If the human body is kept clean, it does not as badly contaminate the objects that it contacts daily. If we remove the obvious soil from the skin, we may still not be clean. The skin of humans and other animals is an excellent site for the growth of bacteria. Secretions on the skin surface provide the necessary nutrients for bacterial growth, and the temperature of the healthy human body is near the optimum for many bacteria. It is essential, then, that we do more than keep ourselves clean of obvious soil by bathing frequently enough to regularly reduce the number of bacteria on the body surface.

The hands perform most of the contact actions of the human with either objects or other humans, and are therefore most frequently contaminated by obvious soil, filth, bacteria, fungi, and viruses. Each time the hands touch an object, microorganisms are deposited on that object, and also are picked up from that object. Therefore, the hands should be washed frequently during the day, and most certainly should be washed thoroughly before critical activities such as handling foods, preparing foods, or eating. The handwashing routine should include adequate washing following each visit to the toilet, and following each incidence of handling any potentially dirty object. In washing the hands, it is important to realize that water and soap alone do not do an adequate job of cleaning. It is also necessary that there be a sufficient amount of rubbing the hands together while they are in contact with the soap and water. The rubbing may remove more microorganisms from the skin surface than do incidental contact with the soap and water. This practice of regular, proper handwashing will reduce the transmission of microorganisms from you to others, and from others and from objects to you.

If individuals practice personal hygiene and cleanliness, then the physical surroundings in the home will be more easily maintained in a sanitary condition. General considerations for sanitation require order and uncluttered surroundings to permit maintaining the house and furnishings with as little dust, dirt, and trash as possible. To keep a home completely free of such is not possible, but general neatness, and frequent dusting and sweeping are sufficient.

In bathrooms, laundry rooms, kitchens, and dining rooms, more specific sanitary precautionary practices are desirable. In bathrooms and laundry rooms, extra efforts are frequently necessary to reduce moisture and to keep the entire space as dry as possible. Not only will moisture permit mildew and produce a musty odor, it will enhance the growth of other fungi and bacteria that may cause human diseases or other undesirable consequences. The high moisture characteristics of these spaces make necessary the use of disinfectants on a regular basis. Special attention must be given to cleaning the toilet bowls, wash basins, rinse basins, bathtubs, and showers. Under closed lavatories, the high moisture content and reduced air circulation make ideal breeding places for cockroaches as well as bacteria (see Figure 1.1). Regular treatment and cleaning of these spaces to prevent insect breeding and growth of bacteria is essential.

In dining rooms and kitchens, it is necessary to keep the premises harmless to human health, and provide for the preparation and service of uncontaminated food. The general rules of cleanliness and uncluttered space apply here, as well as additional care to keep food particles and remains from being left for any length of time. Leftover foods and

Fig. 1.1. Clutter and moisture under sinks attract insects and rodents and allow growth of microorganisms.

food particles are sites for bacterial growth, and also may attract insects. In the kitchen, the same attention must be given to the sink and areas under the sink that is required in the bathroom lavatory. To prevent bacterial and fungal growth, and to avoid attracting and breeding insects, disinfectants and efficient cleansers should be used on a regular and consistent basis.

Appliances should always be left clean; food particles should not be allowed to remain and grease should not be allowed to build up on mixers, toasters, can openers, blenders, food grinders, ventilation hoods, and around cooking burners. One of the pieces of equipment most frequently neglected in cleaning is the can opener, particularly the blade. With heat and moisture available, these serve as excellent incubators for bacterial and fungal growth, and once contaminated with a disease-producing organism, they can allow multiplication and serve as instruments of dissemination of potentially harmful bacteria to many foods or containers in succession. Disinfectants should be used carefully on can openers; otherwise a potentially harmful chemical residue may be left to contaminate foods.

The refrigerator, an appliance designed to save food, must receive special attention in order that it not be a place for dissemination of contamination to foods. The low temperature of the refrigerator, when it is operating properly, will slow the growth of microorganisms, but this is only a process of slowing, and these living cells are not killed. The moisture present, even with low temperature, permits many bacteria and fungi to continue growth at a very slow rate. Unless there is regular and systematic cleaning, with disinfectants, the shelves, walls, and crevices in the refrigerator become heavily contaminated with bacterial and fungal growth. Any spilled food, not removed, serves as an excellent source of nutrient for this growth.

A likely trouble area in kitchen sanitation is the system for removal of garbage and trash. These must be removed regularly and without long delays. The refuse from food preparation, leftover foods, scraps, and soiled wrapping materials and containers will be subjected very quickly to microbial decomposition and may attract insects and rodents. If the waste was contaminated with disease-producing organisms, these will be allowed to multiply and may be spread by insects and rodents to contaminate other foods.

The sanitary practices needed in the kitchen and eating spaces of the home are the same as those required by law in commercial food processing or service establishments, just as the practices needed in the remainder of the home are the same as those required in any public building. Only the mechanics and scheduling of application of the practices really need be changed.

SCHOOL SANITATION

As in the home and public buildings of any type, sanitation in schools, at all levels, is best practiced when there is ample space for the number of persons who use the building. In classrooms and offices, sanitary practices must include methods for keeping down clutter, removal of dust and trash, and general cleaning of all surfaces. Also, as in any group situation, the personal hygiene of the individuals will carry over to the sanitary practices needed, and in actual fact, to the effectiveness of those practices.

Restrooms, locker rooms, and shower rooms must be given additional attention. The numbers of individuals from varied home environments using these facilities daily make it necessary that we assume contamination as a fact and that frequent, regular, and thorough measures be taken to remove that contamination. Cleaning and disinfection must be daily measures in these places to prevent the buildup of disease-causing and other microorganisms. These spaces must have adequate ventilation and dehumidification facilities to hold down microbial growth. The regular use of good, adequately concentrated disinfectants and sanitizers are an absolute must. Foot baths in shower rooms are worthwhile as additional measures for control of athlete's foot and other types of infection; but without adequate frequent disinfection of all floors with chemicals that leave an active residual, these baths are not completely effective.

Lockers can only be as clean as the clothing and equipment kept in them. Athletic clothing and equipment must be cleaned and disinfected before use by different individuals. A regular schedule of frequent cleaning and disinfection will reduce the chances for infectious disease transmission and will greatly aid the sanitary state of the room.

HOSPITAL SANITATION

It is estimated that approximately five percent of patients who are admitted to hospitals for acute care will become infected by organisms encountered for the first time by the patient in that hospital environment. When this amount of infection occurs in patients who are already acutely ill, it becomes one of the most pressing public health problems in the world today. At one time, hospitals were considered the last resort, and in actual practice, patients went there to die. This situation has changed, and hospitals are perhaps used too often now, and this use is one of the reasons for the all-too-frequent occurrence of hospital-acquired or nosocomial infections. The rapid exchange of patients in the hospital environment creates a possibility of rapid exchange of pathogenic bacterial populations. When these organisms reach susceptible individuals, serious infections occur, and for this reason it is essential that strict adherence to schedules of optimum sanitary practices be observed. It is not sufficient that hospital environments be kept free of clutter and dust. The hospital floors, walls, and furnishings must be kept disinfected, particularly when there is a change of occupancy in a hospital room or space.

Effective and sensitive surveillance programs to meet the needs of the individual hospitals are necessary to obtain data on nosocomial infections from which control and prevention programs may be developed. In most hospitals, this surveillance program, as well as the control and prevention programs, are under the supervision of one responsible and well-trained individual, who serves on the infection control committee with representatives of all the critical divisions of the hospital. This committee then sets up the procedures required in each division of the hospital to maintain the necessary sanitary conditions that will prevent the occurrence of hospital-acquired infections. These procedures require that all spaces be not just cleaned, but disinfected, each time a patient leaves, before the next patient comes in. These procedures also require that all furnishings, and particularly those linens and other items that have been used intimately by the patient, be cleaned and disinfected, at the least, and sterilization for most of these items is much preferred. This type of sanitary control is simply an emphasis on the ordinary types of infection control procedures that should be used in any environment to prevent the spread of infectious disease. In the hospital environment where elderly and very sick patients are being treated, there are many other areas that have to be carefully controlled in order to prevent the spread of infection. The increasing use of antibiotics, diagnostic procedures, and therapeutic devices have actually increased the chance that nosocomial infections will occur. These practices involve an increased use of intravenous solutions, transducers, intravascular catheters, inhalation therapy, and controlled air environments. The use of such devices and equipment in patients of advanced age with severe chronic diseases increases the chances for transfer of potentially pathogenic bacteria to these most susceptible patients. In the early days when we became aware of the increased danger of hospital-acquired infection, the most common causative organisms were staphylococci and streptococci. However, changes in medical practices and the changes in the use of hospital care have coincided with the appearance of the more difficult-to-treat gram negative bacilli, and some varieties of fungi, as the organisms most likely to cause hospital-acquired infections. The more frequent diagnosis of nosocomial viral infection in current hospital experience probably only indicates improved methods of viral infection diagnosis.

The surveillance program in hospital infection control is simply a means for responsible hospital personnel to record the occurrence of all potential problems, and to deduce from these the sources of trouble insofar as hospital infections are concerned. Data collected in such a program, if, for example, the inhalation therapy equipment was not kept properly sterilized, would point out the occurrence of infections in those patients treated by use of this equipment, and the condition could be corrected more readily. Records of this type are essential for the hospital, and most certainly for the patients involved. It has been estimated that nosocomial infections cost approximately $2.4 billion annually. Because many of the persons who suffer from nosocomial infections are the elderly who are assisted on Medicare-Medicaid programs, much of this cost is borne by U.S. taxpayers. Any reduction in hospital-acquired infections, either by improved surveillance and control programs for the most common types of nosocomial infections, or from improved routine sanitary and cleaning procedures, will be of general benefit to the economics and the health of the United States.

Prevention of foodborne disease in the hospital environment is essential. Of all establishments in food service operations, the hospital kitchen and dining rooms should be the most carefully controlled from a sanitary standpoint. Strict adherence to handwashing routines, surface and equipment cleaning and disinfecting routines, and proper waste storage and disposal are absolute musts in these areas. Most careful observance of proper temperature controls for holding, storing, or cooking foods is essential, and the use of good quality, fresh foods is another requirement.

The removal of waste and trash from a hospital environment is another area requiring special handling. All trash coming from the hospital must be considered to be contaminated by disease-causing organisms and must be handled accordingly. It is best to keep the trash from patient rooms separate from that coming from other areas. Refuse from surgical suites is always considered to be potentially dangerous, and should be sterilized before being discarded. Whereas in the home, the most undesirable trash may come from the kitchen, in the hospital, the trash most likely to be contaminated with potentially dangerous organisms is that coming from the surgical area, and from the patient areas.

Equipment used in all areas of the hospital environment should be sterilized between uses, as well as laundry used in patient or any other areas of the hospital. Disposable items that can be used in handling patients are desirable, and certainly the arrangements of patient facilities should be such that nurses and physicians are encouraged to wash hands in transit between patients.

SANITATION IN PUBLIC BUILDINGS

Sanitary requirements in public buildings are generally the same as those needed in schools and homes. One difference is apparent: many individuals may not be as personally concerned with the use of such buildings and may, therefore, be less careful about helping to maintain the atmosphere of the building. For this reason, such public places require more attention from a permanent maintenance sanitation staff than would be needed otherwise. Again, the major considerations in maintaining sanitary conditions are adequate space; removal of clutter; regular emptying, cleaning, and exchanging of trash receptacles; and frequent cleaning and disinfection of restrooms, with adequate supplies of bathroom tissue and disposable paper towels at all times. The regularity of cleaning and supplying are critical in keeping these public places sanitary.

OUTDOOR SANITATION

With increasing population size; urbanization; and manufacturing and processing of goods for the use, consumption, and convenience of more and more people; we have observed the need for increased attention to sanitary practices in the outdoor environment as well as in our homes, schools, and public buildings. Environmental concerns, many of which have a basic sanitation basis, involve the production of gaseous, liquid, or solid wastes. Gaseous and liquid wastes present disposal problems that vary according to the waste's toxicity, production location, volume, and site of disposal. These problems are generally handled individually and specifically and are regulated by various national, state, and local agencies. These problems involve many elements of sanitation, but the necessity for outdoor sanitary practices is most frequently seen in the form of solid wastes. A huge proportion of the world's immense solid waste problem is due to the wastes resulting from daily life and activities. It has been estimated that there is an increase of up to two percent per year in the weight of solid wastes resulting from the activities of each individual. It has been estimated that in 1967 in an urban area of an industrialized nation, each person produced 700 kg per year of solid waste. At a two percent annual increase, in 1985 this figure becomes 1,000 kg per year of solid waste. If one compares this to a familiar object, two persons in one year produce solid waste weighing more than a standard mid-sized automobile. Disposal of this waste creates tremendous problems involving sanitation, beauty of the surroundings, and the ecological balance and well-being of the world in which we live. The land, water, and air around us are generally clean, uncluttered, and spacious until man comes in to crowd, spoil, clutter, and desecrate with careless activities and wastes.

Problem areas that are all too apparent can be attributed in large measure to the increase in number of humans, yet we cannot attribute all of our problems just to increased populations. In the United States, we must blame some of our solid waste problems on the fact that we insist upon the convenience of packaging in excess of what is really necessary for provision of most goods. Throughout history and certainly in many parts of the world today, populations are more crowded than in the United States and they do not have the same urgent air, water, and land pollution problems that we face in this country. Many aspects of our way of life have brought us to the point of facing this critical question: How do we practice sanitation in the world at large? For many years we have improved sanitation in our homes, food industries, schools, and public buildings by removal of wastes to the outdoors with little thought of what we were doing in the environment. We have made our land and waters insanitary to the point that we must now learn to practice sanitation there as well. It is in these areas that sanitation and pollution of our environment come face to face as major problems.

SANITATION VERSUS POLLUTION

We have defined sanitation as an attempt by humans to manipulate the environment to their benefit; to manage the environment in such a way as to improve, maintain, or restore good health for humans. Pollution results when people use the environment for immediate convenience without thought of the long-range consequences. One wood fire will pollute the air with a certain amount of smoke. This smoke is rapidly dissipated by simple diffusion and essentially disappears. When this is compounded by millions, and we add to it coal smoke, oil and gas smoke, automobile exhausts, and volatile chemical fumes, we have a major air pollution problem.

One waste can will not create any great problem and will eventually be broken down by natural forces to its basic components, which will return to the soil. Add to this billions of aluminum, glass, and plastic containers that are not broken down by natural cycles of activity, and we have a major solid waste problem.

The municipal wastes from Dubuque, Iowa will do little or no harm to the Mississippi River a few miles downstream. When wastes from dozens of cities, both larger and smaller, are added to those of Dubuque, the Mississippi becomes polluted.

Chemical, solid, and cellular pollutants in the air are harmful to human health. To provide air sanitation, these must be kept from entering the air in harmful quantities. It is, therefore, necessary that persons responsible for the production of these substances inconvenience themselves by reducing pollution for the common good of all. It is to this end that laws and regulations are enacted to govern the release of waste materials to the air, water, or soil by industries or individuals.

Current life-styles in the United States depend more and more heavily on all types of containers and packaging for convenience. In many cases, these containers greatly improve the ease of maintaining sanitary standards, and these standards are improved by the fact that these containers are inexpensive throw-aways. Yet we may find it necessary to face some inconvenience by using containers that can be sterilized and reused, or at least, recycled, unless we can find an answer to the problem of disposal of many types of solid wastes that are not readily degraded and returned to natural components. We have found means of treating household sewage with chlorine, which greatly reduces the chance of disease organisms entering a river in municipal wastes. However, we still have a major problem of solids and chemicals in these wastes that pollute our streams and lakes, and eventually do harm by making water unfit for aquatic life, and unusable as a source of water for human use (see Figure 1.2).

In all areas of environmental pollution, it is essential that we find and practice those sanitary measures that will make it possible for us to dispose of our wastes without harming the world in which we live. Pollution harms all living organisms, and produces an imbalance in nature that may eventually make human survival impossible. Of more immediate concern, the effect of pollution is to increase the dangers of disease in humans by contamination of our land, water, air, and food with harmful chemicals and disease-producing microorganisms. It is necessary that industry, government, and the public practice good environmental sanitation to prevent pollution and thereby to improve the public health.

SANITATION AND BIOLOGICAL SCIENCES

Sanitation, by virtue of its involvement with the protection of the human health, must be concerned with all those segments of the environment that can affect health. It follows, then, that sanitary science must deal with the physical, chemical, and biological factors that make up the environment. Generally speaking, the physical and chemical environments can be managed fairly easily because, without human

Potter (1978)

Fig. 1.2.   Lagoon receiving treated wastewater.

interference, these factors are generally predictable with some degree of accuracy. On the contrary, the biological components of the environment are not always predictable with the same degree of confidence because living organisms react and respond to different physical and chemical stimuli, as well as to other living organisms, including humans. With this potential for response and reaction, it is necessary that anyone concerned with sanitary science have a rather thorough understanding of the behavior of at least those living organisms that are most likely to produce a direct effect on human health.

Biology is the science of life, including all of the specialized branches of sciences dealing with any aspect of living organisms. The extreme breadth of biological science makes it impossible for any one person to be expert in the total discipline, and therefore studies are generally broken down to include animals, plants, or microorganisms in some special grouping. Sanitation must deal with all three categories of living things in that animals and plants serve as food for man; and as will be discussed in later sections, microorganisms are both used in preparation of foods and are also concerned with food spoilage and cause diseases in humans.

Plants are susceptible to disease produced by microorganisms just as are humans. Plants will pick up, and in some cases, concentrate chemicals that may be harmful to man. Plants produce chemicals that may be either beneficial or harmful to man. Sanitation must be concerned with plant biology in order to protect man from possible harmful effects, and to make the best use of potentially beneficial aspects of those plants that grow in our environment.

Animals are susceptible to diseases produced by microorganisms. Many of those diseases may be transmitted to humans either by the living animal, or in food from a diseased animal. Therefore, sanitation must be concerned with both animal and with human health. Smaller animals, including insects, carry diseases to man, to other animals, and to plants. Some single-celled and small multicelled animals produce diseases in humans, in other animals, and in plants. An understanding of the interrelationships that have the potential to affect human health is therefore essential to the practice of sanitation.

Because microorganisms are capable of spoiling food and of producing diseases in man, other animals, and plants, microbiology plays a very important role in sanitary science. To a large degree, the involvement of other animals and of plants in sanitation is a matter of the involvement of microorganisms because of the ubiquitous nature of bacteria and fungi. These microorganisms are found in any environment that has not been deliberately sterilized, and the activities of many of these organisms result in producing the "filthy, putrid, or decomposed" conditions prohibited by the Food, Drug, and Cosmetic Act. The mere presence of many types of microorganisms in an environment constitutes dangerous contamination, although other types that may be present are either harmless or beneficial.

It is essential, therefore, that those concerned with sanitation in the food industries, have a good basic understanding of microorganisms and the relationship of control of microorganisms as these relate to the human. These topics are dealt with in succeeding chapters.

BIBLIOGRAPHY

BOND, R. G., ed.   1973.   *Environmental health and safety in health-care facilities*. New York: Macmillan Publishing Company, Inc.

BRACHMAN, P. S.   1981.   Nosocomial infection control: An overview. *Reviews of Infectious Diseases* 3:640–648.

HOBBS, B. C.   1974.   *Food poisoning and food hygiene*. 3rd ed. London: Edward Arnold Publishers, Ltd.

JAWETZ, E., MELNICK, J. L., and ADELBERG, E. A.   1984.   *Review of medical microbiology*. 16th ed. Los Altos, CA: Lange Medical Publications.

MAURER, I. M.   1978.   *Hospital hygiene*. 2nd ed. London: Edward Arnold Publishers, Ltd.

NATIONAL ACADEMY OF SCIENCES.   1980.   Drinking water and health. Vol. 2. Safe Drinking Water Committee, National Academy of Sciences, Washington, DC.

ROBERTS, H. R., ed.   1981.   *Food safety*. New York: John Wiley and Sons.

TARTAKOW, I. J., and VORPERIAN, J. H.   1981.   *Foodborne and waterborne diseases*. Westport, CT: AVI Publishing Company, Inc.

U.S. DEPARTMENT OF HEALTH, EDUCATION, AND WELFARE.   *Federal Food, Drug, and Cosmetic Act, as amended* in 1965. Washington, DC: U.S. Government Printing Office.

U.S. DEPARTMENT OF HEALTH, EDUCATION, AND WELFARE.   1968.   *General principles of food sanitation*. Food and Drug Administration Publication no. 16.

2

Interactions between Microorganisms and Humans

It is sometimes difficult to keep in mind that we are surrounded by populations of living organisms that cannot be seen with the naked eye. Yet that is our relationship to the world of microorganisms (see Figure 2.1). As was pointed out in the preceding chapter, in order to understand and appreciate the need for sanitary practice, it is necessary to have a basic understanding of microorganisms in general.

To attain this general understanding, we need to know more about each of the kinds of microorganisms. Microbiology is the science dealing with the lives of organisms that consist of a single cell, either complete or incomplete, and require the use of a microscope for study. Most commonly included in this science are: (1) viruses, (2) chlamydia, (3) rickettsia, (4) mycoplasma, (5) bacteria, (6) fungi, (7) protozoa, and (8) algae. Included are two forms that for one reason or another have been reclassified, and placed in separate groups.

VIRUSES

A virus is considered a microorganism but is very different from all other cellular forms of life. It is a submicroscopic particle composed of nucleic acid within a protein coat, and it is capable of infecting a specific living cell and increasing in numbers within that living cell. The virus is said to be filterable, meaning that the virus particle will pass through a filter that will hold back the smallest of bacterial cells. The virus particle is submicroscopic in that it is not visible even with the use of the best ordinary light microscope. The virus can be made visible by use of the electron microscope, which can photograph a specially treated preparation of virus particles. Using the magnifying power of the elec-

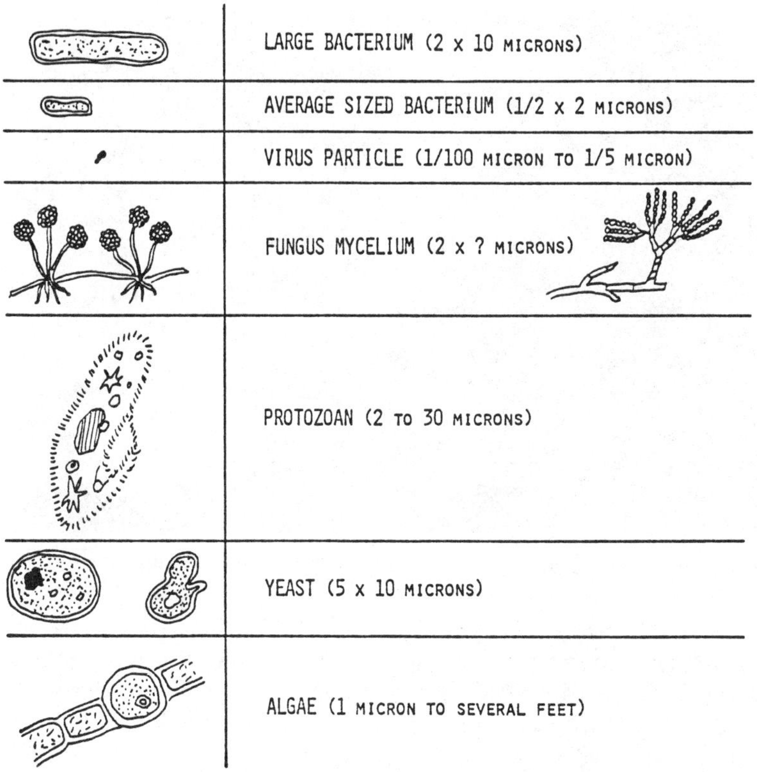

	LARGE BACTERIUM (2 x 10 MICRONS)
	AVERAGE SIZED BACTERIUM (1/2 x 2 MICRONS)
	VIRUS PARTICLE (1/100 MICRON TO 1/5 MICRON)
	FUNGUS MYCELIUM (2 x ? MICRONS)
	PROTOZOAN (2 TO 30 MICRONS)
	YEAST (5 x 10 MICRONS)
	ALGAE (1 MICRON TO SEVERAL FEET)

Courtesy of Robert N. Moore

Fig. 2.1. Microorganisms and their sizes.

tron scope, the size of the virus particle may be determined, along with the shape and at least part of the particle's structure (see Figure 2.2).

The virus particle is not a complete living cell, although it can infect cells. The virus particle does not possess metabolic function, and instead of saying that it reproduces, we say that the particle is replicated by the cell. The particle simply does not have room for all the subcellular structure that we associate with a living cell, but it does contain specific genetic material, and the new virus particles replicated by the infected cell have identical composition to the infecting particle in most circumstances. The virus particle has no means of motility, because it does not possess metabolic function for the release of stored energy; but as fluids are moved in tissues or cells, the virus is carried from one cell to another, and is able to infect specific types of cells. Because

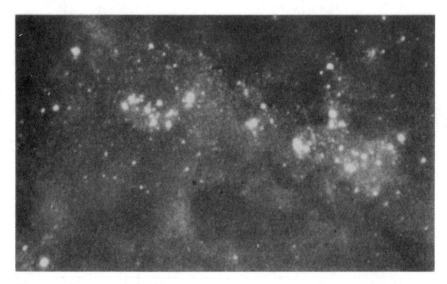

Fig. 2.2. Stained virus particles in tissue cells at a magnification of 10,000 ×.

virus particles cannot replicate outside the infected cell, we do not have to be concerned with multiplication of health-threatening virus particles outside the human body in such environments as food and water.

Viruses are very specific for the type of cell that can be infected, and this specificity does not change. Therefore, most often, human virus infection can be expected to occur only as a result of contact with another infected human, although there are a few virus diseases that are specific for more than one animal species. There are also viruses that can infect only plant cells, bacterial cells, or specific species of animals. For our purposes, we will consider only those viruses that are specific for human cells. This may also include viruses that can infect cells of more than one kind of animal. For example, the rabies virus can infect the nerve tissues of a wide variety of animals, from the bat, to the dog, to man. Others, however, can infect only the cells of one animal species, and in some cases, only one kind of cell within that animal. It is thought that the hepatitis virus A is of this type, infecting only human liver cells. A frequently used example is that of the smallpox virus that infects only the human, and for this reason, when we rid the world of all human cases of smallpox, we, for all practical purposes, rid the world of smallpox.

When a virus particle infects a cell, the nucleic acid portion of the virus actually penetrates the cell membrane and enters the cell to become an intracellular parasite. After this penetration of the cell membrane, the genetic material of the virus particle essentially takes over the operation of the cell, causing it to use its own energy for the manufacture of new virus particles. As new virus particles are completed, one of three different fates may await the cell, depending upon the nature of the virus particle, and certain other conditions within the cell that are not completely understood:

(1) The cell may be destroyed, or lysed, with breakage of the cell membrane and release of the new virus particles to be spread to other susceptible cells.

(2) Other types of viruses will stimulate the cell to increase in size and number. In this case, some cells may be destroyed and new virus particles are released to be spread to other susceptible cells.

(3) In some cases, the virus may simply be released through the cell membrane leaving the cell essentially unharmed, except for the receptor, which was destroyed at the time of infection. When this receptor is rebuilt, the cell may again be infected by specific virus.

Whichever type of infection occurs, there is some disturbance in the normal structure and function of the cells, and of the human so infected, resulting in a disease process.

Occasionally a very different series of events occurs after a virus infection. The virus infects in the usual manner, but once the cell is infected, the viral nucleic acid material remains latent, or inactive, and does not stimulate the cell to produce new viral particles. Instead, the latent virus may remain within the cell, acting as a natural part of the cell, and when cell reproduction takes place, the viral nucleic acid is also reproduced, and becomes a part of the new cell. The virus may remain latent for an indefinite period of time, but eventually, something disturbs the cell and the latent virus becomes active, stimulating the production of new virus particles. As long as the virus remains latent, no disease process occurs, but when new virus particles are produced, the new virus particles are released as described earlier, and the symptoms of the infection appear. A new virus infection can be transmitted from one individual to another only when new virus particles are being produced by and released from infected cells.

CHLAMYDIA

Chlamydia are a group of obligate intracellular parasites that possess many, but not all of the characteristics of some of the smaller bacteria. These organisms contain nucleic acids, and they multiply by binary fission. They have a cell wall and a limited variety of metabolic enzymes. Their growth can be inhibited by some antibiotics. Chlamydia, however, cannot carry out sufficient independent metabolism to multiply outside living cells. These organisms are very sensitive to heat, completely losing the ability to infect cells after a very few minutes at 60° C. The chlamydia may very well reach a balance with the infected host cells, and thereby produce a very prolonged infection or co-existence with the host. Some species of chlamydia frequently spread through direct contact between humans to produce disease; but with other species, the human is more often infected by organisms from another animal species (frequently birds).

RICKETTSIA

Rickettsia are also intracellular parasites. In this they are very similar to the viruses, although they differ in many other respects. The rickettsia are not able to pass through bacterial filters, and are large enough to be just visible with the use of a very good light microscope. With the larger size of the rickettsia, there is also a greater amount of internal structure and organization, and there are more activities that correspond to the functions of the typical living cell. Rickettsia, however, are not able to reproduce outside another living cell.

Rickettsia normally parasitize fleas, mites, lice, and ticks and are transmitted to people by the bite of one of these animals. Many rickettsia are able to reproduce in human cells and when this occurs, disease results. Rickettsial diseases are not normally spread from person to person and are therefore best controlled by control of the animal vectors, or carriers of the organisms. At present, rickettsial diseases are not extremely common in this country; however, breakdown of sanitation to the point where animal vectors are not controlled could result in a much greater incidence of these diseases.

MYCOPLASMAS

Mycoplasmas were at one time called pleuropneumonialike organisms (PPLO) and are different enough from both bacteria and other microorganisms to deserve some special mention here. They lack any rigidity in the cell wall, but are surrounded by a triple-layered membrane. They are complete living units in that they will grow on nonliving culture medium, and are capable of reproducing without other living cells. These organisms are not bacteria that have simply lost their cell walls, but are separate, independent living organisms. Their growth is inhibited by many antibiotic compounds, and by specific immune antibodies produced in infected animals. In many characteristics, these organisms are similar to the chlamydia described in an earlier section, but are able to reproduce independently.

Mycoplasmas that cause disease appear to be very host-specific; that is, they will infect only one species of animals or plants. Within that species of animal or plant, however, the organisms are communicable. In animals, the organisms appear to be intracellular parasites, with transmission of the organism by respiratory discharge or by direct contact.

BACTERIA

Bacteria are single-celled living organisms that possess a wide variety of characteristics, and in many cases an amazing ability to adapt to adverse conditions in their environment. They may be observed and studied with the use of an ordinary light microscope (see Figure 2.3), and when grown in culture, groups of cells, or colonies, are visible to the naked eye. Unlike viruses and rickettsia, most bacteria are saprophytic, not parasitic. As saprophytes, they use nonliving nutrients and exist in a free-living environment. Parasitic bacteria are almost always extracellular parasites, although a few species may at times actually invade the cells of an infected host animal or human.

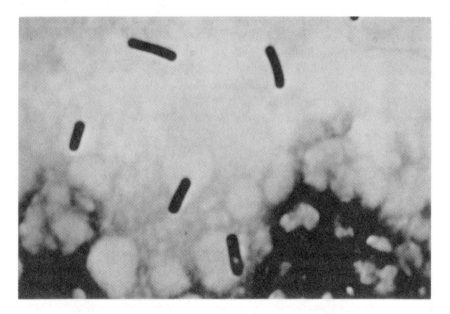

Fig. 2.3.   *Clostridum perfringens* cells at a magnification of 5,000 ×.

Even though many different bacterial species are saprophytic, they adapt to live in or on the animal or human body. Most of those that can adapt in this manner are included in the group called "normal flora" and many of these will cause disease only in unusual circumstances. Others are opportunists and will be more likely to cause disease, while some are considered to be pathogenic and should be avoided to ensure disease prevention. A few species of bacteria adapt to live in or on the human body and benefit the individual by supplying needed vitamins, by helping to control the water balance of the intestinal tract, or by competition with pathogenic bacteria to aid in resisting infection. These organisms will be discussed in greater detail in later sections.

FUNGI

The fungi are a diverse group of plantlike organisms and microorganisms that can be considered in two groups: the yeasts and the molds. All fungi are organisms that do not have specialized tissues such as roots, stems, and leaves. Fungi do not have chlorophyll; instead they obtain their food from nonliving organic matter or by feeding as parasites on living hosts.

Sac fungi (yeasts) have been known and used by humans since long before the yeast cell could be observed. Yeasts were first used for fermenting fruit juices, leavening bread, and making certain foods more palatable. Very few yeasts are dangerous to the human, and most of them are useful. Yeasts are always single-celled organisms and may reproduce by budding, cell division, or spore formation (see Figure 2.4). Generally speaking, the yeasts may be divided into true and false yeasts. True yeasts are those used in baking and brewing, while false yeasts may be of medical importance in some cases.

The algaelike fungi (molds), unlike the yeasts, do not always exist as single cells. Molds form long filaments called hyphae. In many cases, these filaments extend into the air and appear as cottony, velvety masses. Most molds are saprophytes; some are beneficial to humans. One, *Penicillium notatum,* produces an antibiotic, penicillin. Antibiotic substances are chemicals produced by one organism that are toxic or inhibitory to other organisms. Many of these are useful in treating diseases, and have been called wonder drugs because of this usefulness. Some other types of fungi are used in the production of food, providing excellent sources of protein and carbohydrates. Both molds and yeasts are very good sources of vitamins. Many molds are used in the manufacture of cheeses; the type of mold that is used is actually responsible for the flavor of the cheese. One group of molds, the *Fungi Imperfecti,* contain most of the fungi that cause human disease. Many other fungi

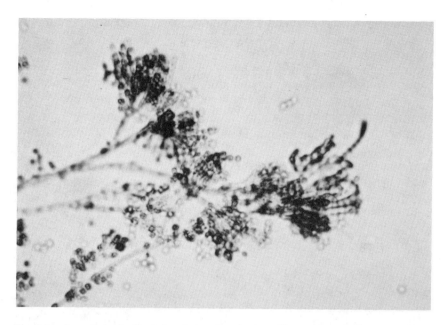

Fig. 2.4. A portion of a fungal mat showing the long filaments and spherical spores.

indirectly do harm by causing the decay of wood and the breakdown of textiles, paint, and so on. Mushrooms are fungi; some of these organisms are good sources of food, and others are extremely poisonous.

PROTOZOA

The protozoa are microscopic animal-like organisms. Most protozoa are free-living saprophytic forms that affect the human only indirectly as members of the food chains in nature. All are single-celled; all have the typical characteristics of an animal cell (see Figure 2.5). A few of the protozoa are parasites that live in human or other animals tissues and produce disease. The protozoa, although similar to animals, are included with the microorganisms because of the single-celled characteristic and because a microscope is required for their study.

ALGAE

This group of microorganisms is made up of single or multicellular plantlike organisms. They are found most often in water, although a few species have adapted as soil inhabitants. Most algae are not harmful and, in fact, can be considered to be at least indirectly beneficial to

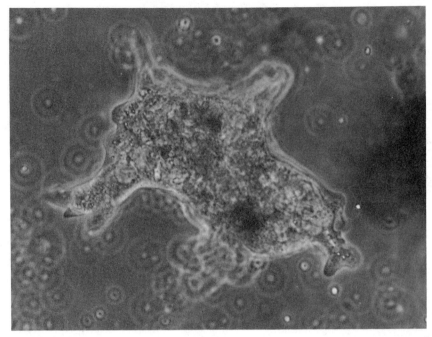

Courtesy of J. K. Reed

Fig. 2.5. A protozoan of the *Amoeba* genus at a magnification of approximately 250 ×.

humans. Algae, like plants, do possess chlorophyll, but they do not have specialized structures such as roots, stems, and leaves, although some of them grow to considerable sizes.

NORMAL FLORA, OPPORTUNISTS, PATHOGENS

Of the tremendous numbers of microorganisms recognized, only a few of them are considered to be pathogenic to the human, that is, these microorganisms will usually cause infection when transmitted to a location from which they can gain entry into the human or animal body. A microorganism can produce disease only when it manages to reach this special location, the portal of entry. It must then enter the body, travel to its preferred location, and increase in numbers. While this is in progress, the microorganisms are also frequently producing a variety of chemicals that interfere in some way with the normal functioning of body tissue cells, a condition that we call disease.

Most microorganisms are saprophytic; that is, free-living outside the human or animal body and utilizing nonliving substances for food. Of such saprophytic organisms, many are actually beneficial in that they maintain a balance in nature, produce chemicals antagonistic to disease-causing organisms, or produce a wide variety of chemicals useful to man (i.e., vitamins, enzymes, proteins for food supplements, etc.). Because of these benefits, control measures aimed at preventing infections must not be so all-inclusive as to eliminate beneficial organisms or even to alter the natural balance of the microbial communities surrounding us.

A group of organisms of concern to humans, and to sanitation science are the heterogeneous group termed *normal flora*. The definition of this term must be limited, and there must be no use of the term to include organisms outside those limits. The microorganisms consistently present in or on the human body in ordinary circumstances of good health are called normal flora. The normal flora of one individual will differ, sometimes rather markedly, from those on another, but under stable conditions, the normal flora of one individual will remain rather constant over relatively long periods of time. Organisms with the normal flora may include

(1)   some saprophytes that have accidentally lodged with the human and adapted to the environment of temperature, moisture, and nutrients present;
(2)   organisms that prefer the environment of the human body for this environment; or
(3)   potentially pathogenic organisms that do not survive well outside the human body.

In this environment, they may flourish if there are few inhibitory factors, but their presence is not absolutely necessary for life, although they may play some role in maintaining the health and normal functions of the human body. In the intestinal tract, some microorganisms, which are a part of the normal flora, synthesize vitamins and aid considerably in the absorption of water and nutrients from the intestinal tract. On skin and membranes, some of the normal flora organisms may help protect against potential pathogens by competition for nutrients, or even by producing chemicals that will inhibit the growth of pathogenic species. When the normal flora of any part of the body are suppressed or removed by some treatment, other organisms will move in to fill the vacancy, and may result in the production of disease.

Organisms that make up the normal flora for an individual do not usually produce infection; however, certain of these microorganisms can do so if provided the proper environment in terms of lack of competition, optimum temperature, moisture, and nutrients. When this occurs, the organisms so involved are termed *opportunists*. Any disturbance of the balanced existence of the normal flora population may result in opportunistic infections. Another event that may result in this kind of disease is the introduction of some new microorganism to the environment.

Under normal circumstances, there is a rather specific group of organisms that can be expected to colonize certain parts of the human body and to comprise the major portion of the normal flora populations. For example, certain bacterial types prefer to inhabit the external skin; others, the mouth and respiratory tract; others, the intestinal tract with its very different pH and anaerobic conditions; others, the urinary tract; and others, the eye. Within these locations, the organisms that are expected to be present do not normally cause any difficulty for the human, but if moved from these to other areas of the body tissues, they may become opportunistic, or even pathogenic.

In rare instances, individuals may possess some pathogenic organisms in their normal flora. In such cases, the pathogen is held in check by the presence and/or activities of the other organisms in the environment. A pathogen is defined simply as capable of causing an infectious disease, and such an organism will ordinarily be able to produce that disease if present in sufficient numbers. Fortunately for humans, however, not every contact with a so-called pathogen results in infection. At times the pathogen is present in numbers too few to initiate the disease process; at other times organisms in the same environment will by some means inhibit the pathogen to prevent establishment of the disease process; and at times the individual providing the environment is simply too resistant in various ways to permit the initiation of the infection.

TRANSMISSION OF INFECTIOUS DISEASE

Infection is the invasion of human body tissue or of the body of other animals or plants by microorganisms with the production of damage. In some cases the damage is slight; in others the damage is extremely severe. In all cases the infected body puts up some defense against the invading organism. Thus, in an infection action is taken by the invaded host and by the invading microorganism. Essentially, infection is the result of a race between the defenses of the infected host, and the invasion or multiplication of the pathogen. If the race is won by the defenses of the host, the infection is overcome and little damage is done. On the other hand, if the race is won by the invasion or multiplication of the pathogen, the infection progresses with more and more tissue being damaged. Because the human body is constantly exposed to all varieties of microorganisms, many of which are capable of producing disease, it is obvious that the defenses of the host win most of the battles in this continuing fight against disease-producing organisms. If this were not the case, all of us would constantly suffer from some disease produced by microorganisms. It is at present impossible for the human body to exist in the complete absence of microorganisms. A few other animals involved in experiments have been raised in an apparent germ-free environment; however, some species tested in this manner have not been capable of surviving without the presence of microorganisms. In one case, a boy survived to the age of 12 years in a germ-free environment—a situation made necessary by the complete lack of any system of resistance in his body. The survival was not terminated because of his failure to survive in the germ-free environment, but rather due to treatment in efforts to permit the individual to live a normal life. Because we must live with the microorganisms, it is necessary that we learn to control those that surround and constantly attempt to invade us. For the purpose of control of these organisms, it is helpful to understand how a chain of infection or the transmission of infection can be interrupted. To do this, we need to understand what the mechanism or what the manner of transmission of disease is as it exists in our environment. Obviously, it is to our advantage to break the chain of transmission without unduly interfering with our normal life-style.

To consider the means of transmission of infectious disease from one person to another, it is necessary to consider both the pathogenic organisms that are expected to produce disease and those organisms termed opportunists that produce disease occasionally. Let us consider the opportunists. Because these microorganisms are somewhat unpredictable in production of disease, it is very difficult to lay down definite, unalterable rules by which we may control the presence and activities of opportunistic microorganisms. Rather than specifically attempting to eliminate any one of these from our environment and thus eliminating the possibility that they may produce infection within us, it is more practical to approach this problem as a matter of general control of all microbial populations. If we are to do this, then we must also take into account the fact that some microorganisms are beneficial to man. Therefore, it is not desirable to eliminate all organisms from our environment; rather, it is desirable to control the numbers of organisms that may be present in our surroundings. In this respect, then, we practice sanitation; controlling the numbers by preventing the multiplication of microorganisms at too fast a rate. This will keep a lower number of microorganisms in our environment and quite likely this lower number will be insufficient to start infection should they happen to gain entry onto or into the human body. For infection to be produced, a certain number of the specific infecting microorganisms must be present at the infected site. This number is referred to as the infective dose. Again we can see the similarity with a battle. If the armed forces are too small in number, it may be impossible to gain a particular objective, and to reach the defensive position of the enemy. Similarly, with the microorganism attempting to invade the human body, one cell attempting to gain entry simply will not have sufficient ammunition to get into the body tissues, to progress, grow, and to do damage. However, if we have several hundred cells at this site, their collective activity may provide a wedge to gain entry into body tissues. The collective reproduction of these cells may produce numbers that will allow the infection to progress. Thus, control of numbers of organisms present in our environment will greatly reduce the chances that we will be infected, particularly when these organisms are only opportunistic and are not particularly adept at producing infection. We may control these numbers by the application of disinfectants to our environment; and keep in mind that we are not always talking of a strong disinfectant such as you might use to scrub a floor. We may be dealing only with ordinary hand soap and water. If good personal hygiene is practiced, using this same disinfectant or soap and water on our bodies, we help to maintain control of numbers of microorganisms and thus help to reduce the possibility of infection.

We must consider that among these microorganisms we are attempting to control are some pathogenic microorganisms quite capable and especially adapted for the production of disease. This adaptation includes special capabilities for overcoming the defenses of the human body, for invading the tissues, and for doing damage. With such pathogenic organisms, if we simply attempt to keep the numbers low, we may not be successful in preventing disease. To combat pathogens, present only in low numbers, we also have the presence of beneficial and opportunistic organisms assisting. Even though we have kept the numbers low, they are still present in or on our bodies, and the presence of these helpful or, at the least, harmless microorganisms will provide competition for growth factors, space, and nutrients that are required for multiplication of the pathogens. This competition actually will benefit by holding down growth of the pathogen and preventing it from gaining in number, which would permit it to advance in its attempt to invade tissues and produce disease. While these organisms are present in or on the human body, the pathogens will never have a completely free reign, although they may be maintained and will retain the potential for producing disease. While pathogens are present even though not doing harm to the individual concerned, there is always a potential that these microorganisms may be transferred to another human and in that person find desirable conditions for progressing to active infection. Each of us in our contact with our neighbors serves as a potential source of infectious organisms, as long as we carry any pathogenic organisms in or on our body.

How are microorganisms, either pathogenic or opportunistic, transferred from ourselves or our environment to the body or the environment of another individual? Such transfers are carried on the "paths of infection," and it is on these paths that we must throw up roadblocks to provide good sanitary conditions (see Figure 2.6). We must come to realize that if an organism is present in our breathing passages or in our mouths, it is also likely to be present on our skin. We live so intimately with our own bodies that we manage to maintain a rather constant reshuffling of all the microorganisms present in or on it. Because microorganisms are present on the skin, one of the first paths of transmission that comes to mind is the direct contact of one human being with another. How often during the day do we shake hands or lay a friendly hand on another person's shoulder? We are constantly touching other humans in our work when we accidentally touch; in

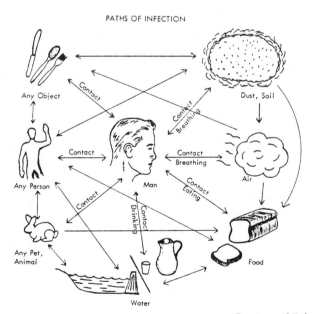

PATHS OF INFECTION

Courtesy of Robert N. Moore

Fig. 2.6. Paths of infection.

physical exercise as we hand one object to another person; and in the more intimate associations of one human with another. This contact provides a direct route of transfer or, rather than transfer, of actual exchange of microorganisms present on bodies, clothing, or objects as we come in direct contact with each other. In most situations, the number of microorganisms exchanged in a brief, chance contact will be very small. In more intimate contact, the numbers will be larger. However, an organism that is an opportunist and is doing no harm in one person may find favorable conditions in the next and then will be able to multiply and cause disease. Therefore, even the transfer of a limited number of organisms in this manner may be sufficient to transfer disease and to spread infection.

The above example is direct transmission of microorganisms between a source and a recipient, sometimes susceptible, host. More indirect transmission, but frequently much more efficient transmission of potential disease organisms from one human to another is seen in two most natural physical acts which all of us have experienced numerous times. These acts are the cough and the sneeze. As mentioned earlier, if we have organisms in any one place on our bodies, the chances are excellent that they will be in some other place as well. It is apparent then, that at one time or another, most of the organisms on our bodies will be present in our mouth and nasal passages. If we sneeze, we spray droplets of water from the mouth and nasal passages out into the air in front of us. How far will these droplets, carrying perhaps thousands to millions of microorganisms, be carried by a good healthy sneeze? In attempting to mimic the actual sneeze it is quite easy to put a small amount of a liquid culture of bacteria into a balloon and inflate it so that the bacteria are present over the entire inner surface. Then, setting culture dishes across the room, we may see how far the bacteria may be carried when the contaminated balloon is punctured. Such tests have demonstrated that bacteria may be spread as far as forty feet from the balloon. Most of the rooms in which we carry on our daily occupations are not forty feet wide, so that obviously with one sneeze we may spread organisms from our mouth, nasal passages, and breathing passages across most of the rooms we occupy. A droplet of water large enough to be visible may contain several million bacterial cells and even more virus particles. Therefore, we are providing sufficient numbers in one droplet from a sneeze or cough to produce an active infection in the person sitting across the room. Regardless of our manners and social graces, most of us are guilty, at one time or another, of sneezing or coughing suddenly before we are able to cover our mouth and nose, and in that way we move microorganisms that may be present in our respiratory tract to those individuals who are present in the room with us. For the transmission of many infections, this one pathway is quite sufficient to keep these diseases going throughout a large population and eventually to bring them back to us.

An additional mode of transmission is more indirect than those just mentioned. This mode may be termed *vehicle-borne* transmission, in that some object carries the microorganisms between individuals. In the preparation of foods, the cook is constantly touching the foods during mixing, handling, and preparing. The cook may sneeze or cough so that microorganisms are transmitted to the exposed foods. If the food is to be cooked following this handling, most disease organisms that were transferred to the food will be killed by the cooking process. If, however, the food has already been cooked, or is not to be cooked, or is to be cooked or warmed only for a very brief time, we have provided an indirect transfer of many microorganisms from one human to another. If the numbers are low and the food is eaten rather soon thereafter, the chance of transfer of an infective dose is reduced. However, as any cook is aware, in the preparation of a meal all dishes cannot be prepared at one time, and therefore one dish prepared and left to sit while other dishes are being prepared provides an opportunity for microorganisms to multiply and for an infectious dose to be produced in the case of bacteria. If this dish is left near the heat, it is kept warm, thereby allowing an excellent opportunity for a few bacterial cells that may have contaminated it to grow into many bacterial cells before the food is consumed. For some common bacterial species, it requires only about fifteen minutes for the population to double, so that if this process is repeated for an hour or more, quite obviously what began as a few bacterial cells is now a sizable population of bacteria. If a few of these were pathogenic, then the population of pathogens would be quite large by the time the food was eaten.

Some foods are contaminated before they reach the kitchen. Vegetables grown in the garden are going to be contaminated by all of the microorganisms in the soil, the air, the fertilizer, and the water. Some of these are brought into the kitchen. Some of these are pathogenic. Some of these are opportunistic and under the right circumstances are capable of causing infection. If the vegetables are cleaned, the numbers are reduced many fold. If the vegetables are not properly cleaned and time is allowed for multiplication of the organisms present, we simply increase the numbers of microorganisms and serve these in our food.

Many foods must be processed before they reach the kitchen. They are processed by other humans who come in contact with the foods and who most certainly contact the instruments and utensils used in the processing. The people who do the processing and the instruments and utensils used in processing all will have some microorganisms present. If a few of these are opportunists or pathogens, they have an opportunity to grow from the time of processing to the time they reach the kitchen, through the time of preparation, to the time of serving. In the consumption of such foods, we may ingest sizeable numbers of microorganisms of either pathogenic, opportunistic, or saprophytic varieties.

The use of water in the preparation of foods will add some microorganisms to the foods, particularly when this water has not been treated for removal of microorganisms. Although the water may have been treated and be perfectly safe for drinking, it may still contain a few microorganisms that in such numbers are harmless, but that when allowed to incubate in prepared food, will grow into potentially dangerous numbers.

Once they are prepared, foods may be contaminated by organisms from the air if left exposed for even brief periods of time. They may be contaminated by insects, which will be attracted by food odors. Such insects are potential carriers of many kinds of microorganisms because they are also attracted by odors of waste products in which pathogenic microorganisms may be present and actively growing. Contamination by insects may be one major means of transmission of disease, because this contamination is most likely to occur after preparation and cooking. At this time the food is warm and provides an excellent environment for the multiplication of disease-causing organisms. This is termed *vector-borne transmission* and the insects that carry the microorganisms are termed *vectors*. If the microorganisms do not multiply in the insect, then the insect serves simply as a mechanical vector and smaller numbers of microorganisms are likely to be transmitted.

An additional pathway of transmission of microorganisms from one human to another is also indirect and it may simply be a contamination of our environment by ourselves. In our daily occupations and activities we are constantly using and touching all sorts of objects. As in the case when we touch another person, we leave some portion of our bacterial flora on the touched object, and we pick up some portion of the bacterial flora that was present on the object. The next person to touch or use that object will then receive the microorganism that we left there. Common objects that are excellent vehicles of transmission for disease organisms are pencils, books, towels, drinking cups, drinking fountains, shower stalls, bathtubs, washbasins, and in athletic environments, the floors where individuals have walked after showering. All of these are potentially dangerous as reservoirs of microorganisms left by some other individual.

CHEMICALS AND DISEASE

Disease is defined as a "condition in which bodily health is impaired; sickness; or illness." Thus, infectious disease is only one type of disease, and certainly health can be impaired other than by microorganisms, as discussed earlier. Also of concern in sanitation are the diseases that may be produced by certain toxic chemicals when these are taken into the body. Many chemical substances are toxic to the human when inhaled, or when contacted externally. Some of these toxic substances are used by man in controlling the environment, and thus the chances of contact are much increased. Microorganisms produce some chemicals that are toxic to humans, and that, if present in food, will cause disease when that food is consumed.

In sanitation practice, it is sometimes necessary that chemical poisons be used to prevent the infestation of buildings by rodents or birds. These poisons are generally toxic to humans as well as to the pests, and great care must be exercised to prevent contamination of foods or other exposure of humans. Insecticides must frequently be used to prevent roaches, ants, flies, and mosquitos from invading homes or public buildings. These chemicals are toxic to humans and must be used with care. The agricultural industries have been able to increase food and fiber production to fantastic levels by proper use of insecticides and other pesticides to prevent crop damage by insects and other animal and plant pests. These chemicals, used properly, benefit humans and the environment; however, used carelessly, they may contaminate food or come in contact with the human in sufficient concentration to cause disease. Other than direct exposure of humans to pesticides during manufacture, handling, or use, both food and water are subject to contamination by these chemicals. The FDA has long enforced limits, or maximum allowances for pesticides on or in food products. The Safe Drinking Water Act of 1974 and its enforcement includes limits on presence of the pesticides that are in general extensive use and that are potential contaminants of drinking water. The National Research Council and National Academy of Sciences selected fifty-five pesticides from this category for detailed study as to potential toxicity for people or animals consuming or using the water.

The safe use of pesticides requires that these chemicals be biodegradable, which means that living organisms must be able to metabolize the compounds. The organisms most often associated with this degradation process are the microorganisms, particularly bacteria and fungi. When these microorganisms metabolize such biodegradable substances, they generally increase in number. The potential for difficulty from this source has not yet been fully assessed for areas where pesticide use is especially heavy.

Many chemical substances are routinely required in industrial and manufacturing operations. Some of these are then disposed of through waste systems which are not always totally efficient. When these substances are in sewage, they contaminate streams and lakes; they may make the water directly dangerous to humans, or may become concentrated in aquatic life to an extent that will endanger persons who consume fish from this water. The EPA has, for several years, been compiling a list of toxic substances that have been or may be released in effluents from industrial plants. The National Research Council, National Academy of Sciences reported that 298 volatile organic compounds have so far been identified in drinking water, and chose 74 of those for detailed study of potential toxicity. As with pesticides, these chemicals must be biodegradable or removable by water treatment to avoid detriment to humans. Toxic chemicals of these types may be easily and beneficially used to the advantage of man when handled properly. Improper and careless use present grave chemical-disease dangers to humans. The precautions necessary to prevent these dangers are integral parts of good sanitary practice.

MICROORGANISMS IMPORTANT TO FOOD PRODUCTION

The use of microorganisms in food production may be traced to the early days of recorded history in bread-making and in the fermentation of juices to make wines. In both of these processes, yeasts produce the desirable chemicals by breakdown of carbohydrate molecules. If other organisms are present, particularly bacteria, the desired chemicals may be further changed and the bread or wine spoiled.

As we came to understand the fermentation process, more information was obtained concerning the chemical activities of other microorganisms, and many of these activities have been put to use by man. The spoilage of wine by bacteria can lead to the production of vinegar. Commercially, vinegar may be produced by allowing fruit juices to be fermented by a yeast, *Saccharomyces cerevisiae,* to produce ethyl alcohol. The ethyl alcohol is, in turn, converted to acetic acid by the activity of bacteria, *Acetobacter* species.

By the production of certain chemicals, many microorganisms cause desirable changes in certain foods. *Leuconostoc mesenteroides* is used to produce sauerkraut and certain types of pickles. Certain fungi produce large amounts of citric acid from carbohydrates. The citric acid is then used for flavoring foods, in candies, and in some soft drinks. *Lactobacillaceae* species of bacteria produce large quantities of lactic acid, and *Streptococcus lactis* is the major milk-souring bacterium, leading to the production of other milk products. *Leuconostoc citrovorum* may be used as a culture in butter production to improve flavor. Fungi of different kinds are used to produce the characteristic flavors of cheeses, for example, blue, Camembert, Roquefort, and Limburger. *Propionibacterium shermanii* changes lactic acid to propionic acid with the production of carbon dioxide gas. This process produces Swiss cheese and the gas forms the characteristic holes in this product.

The chemical activities of other microorganisms are harnessed and utilized to produce characteristic qualities of certain foods, and also to produce such food additives as vitamins and nutritious protein supplements. Science adapted these qualities and abilities of microorganisms to benefit humans.

OTHER BENEFICIAL MICROORGANISMS

The law of conservation of matter states, "matter can neither be created nor destroyed." Therefore, the chemical elements that exist in our world cannot be destroyed, neither can we create new ones to replace the old. Rather, there is a constant change, or recycling, of all chemical elements present. Living organisms of all kinds are necessary for this continuous process and microorganisms of many types play a very important role in some of the cycles necessary for the maintenance of the varieties of life now known.

One of these essential processes is known as the nitrogen cycle, in which atmospheric nitrogen is fixed by certain bacteria into chemical compounds that can then be utilized by some plants (see Figure 2.7). The plants convert these nitrogen-containing chemicals into plant proteins, which are consumed by animals. Animals, in turn, convert the plant proteins to animal proteins. The human benefits by use of both plant and animal proteins. As bacteria, plants, and animals die, the proteins are broken down, in large part by the activities of microorganisms, to simpler chemicals containing nitrogen that can be reused by plants, or that may be further degraded to atmospheric or gaseous nitrogen and the cycle must be repeated. Because plants and animals are unable to utilize gaseous nitrogen, the microorganisms in these processes are essential to the continuity of living organisms.

In like manner, other chemical elements are constantly recycled by decay of dead organisms, conversion of the chemical components to more useable compounds, reuse and restructuring of these food elements, and finally decay and reentry into the cycle. Microorganisms again play leading roles in these processes. Important chemical cycles of this type include the carbon cycle and the sulfur cycle. In this way,

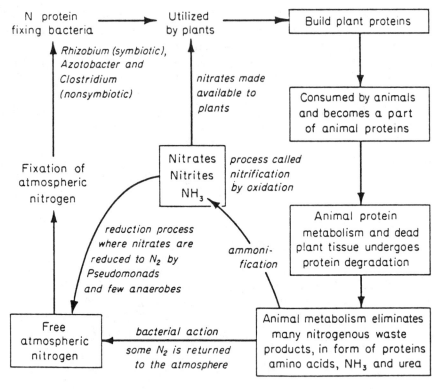

Weiser et al. (1971)

Fig. 2.7. The nitrogen cycle: Protein → peptones → proteoses → polypeptides → peptides → amino acids → NH_3 → NH_2. NH_3 is oxidized to nitrites by *Nitrosomas* bacteria. Nitrites are oxidized to nitrates by *Nitrobacter* organisms. Denitrification is undesirable, because the nitrates are reduced to elemental nitrogen and returned to the atmosphere before they are utilized by the plant.

the fixed amount of food or raw material available to us is kept in service and permits the continuity of life and life processes.

The decomposition role in the cycles just discussed is of major concern in that each type of living organism must be supplied with rather specific forms of chemical compounds. Thus, when a living organism dies, before those structural chemicals can be used by another form of life, they must be degraded and restructured to suit the special requirements of the consumer. Microorganisms of many types, particularly among the fungi and bacteria, are essential to this decomposition and restructuring which must occur before the chemicals can be returned to the food supply of the earth.

Obviously, then, microorganisms are of great benefit to humans and to the environment in many ways. These tiny living forms are efficient chemical laboratories that can manufacture foods, drugs, and other chemicals for humans. It is also true, however, that these very small creatures are, in some cases, harmful to humans, and these potentially dangerous organisms must be understood and controlled. Some of the dangers involving microorganisms as concerns food sanitation are discussed in the following chapter.

BIBLIOGRAPHY

EDMONDS, P. 1978. *Microbiology. An environmental perspective.* New York: Macmillan Publishing Co., Inc.

NATIONAL ACADEMY OF SCIENCES. 1977–1983. *Drinking water and health.* Vols. 1–5. Safe Drinking Water Committee, National Academy of Sciences, Washington, DC.

3

Foodborne Diseases Caused by Microorganisms

Foodborne disease is too frequently considered to mean only those diseases that most often show at least some gastrointestinal tract symptoms. While these diseases are the ones most often proven to be foodborne, they are by no means the only infectious diseases that are transmitted to man by contaminated foods; which is the definition of *foodborne disease*.

Throughout history, the diseases transmitted by foods have not changed appreciably; however, our recognition of the transmission of infectious diseases by contaminated foods *has* changed in recent years. Two possible sources of food contamination exist. The first of these is contamination of the raw food in its original habitat by way of air, fertilization, irrigation, or exposure to the general environment. The second possibility is that people may transfer opportunistic or pathogenic organisms from a carrier or infected individual to the food in the process of cleaning, processing, preparing, cooking, or serving it. Because of these sources of contamination, it is necessary to consider food from the standpoint of preservation; that is, preventing spoilage from natural contamination and preventing additional contamination during the process of preparing the food for consumption.

Food preservation has been a concern of mankind since the beginnings of human history. Long before the causes of food spoilage were understood, it had been learned that food treated in certain ways would not decay or spoil as rapidly, and in some cases the texture and/or flavor of the foods would actually be improved by such treatment. Early attempts at food preservation included the processes of drying and cooling. Dried foods formed a large portion of the diet in the days before modern preservation methods were available. Man learned to dry foods after observing that grains and certain nuts did not spoil in their natural state, in which much of their moisture was lost. By using dehydration, we learned to preserve many foods and more recently the range of foods preserved by this means has increased. One advance in dehydration as a method of food preservation was made in the combination of drying and freezing. By drying from the frozen state, the texture and nutritional quality of certain foods were maintained in a more desirable condition. Dried fruits, vegetables, milk, fish, meats, and eggs are common staples in our present diet, and both the flavor and the nutritional values of these foods are improved by the use of dehydration. Dehydrated food will not spoil as long as it is kept dry. All living cells require moisture to grow and without this moisture, growth cannot occur and some cells die. In addition to the removal of moisture from dehydrated foods, the process increases the relative concentrations of sugars, acids, and inorganic salts that are present as components of the foods being dried. Increasing the concentrations of these chemicals also has preservative action in that the increased concentrations inhibit or retard growth of most living cells.

Rather early in history, we learned that food would keep longer if stored at a cooler temperature. It was placed in shady places, cool breezes, or sometimes submerged in containers in cooling, flowing water. We now substitute refrigeration to help maintain foods in an edible condition for longer periods of time. Microorganisms that contaminate foods and cause spoilage all have a temperature that is most desirable for growth of that particular organism and that permits the most rapid growth (see Table 3.1). Most decay-producing organisms have an optimum temperature somewhere in the range of ordinary room temperature or that of human body temperature. Therefore, if foods can be maintained at a temperature less than this, the growth of microorganisms is retarded and food spoilage does not occur as rapidly (see Figure 3.1). It must be remembered, however, that cooling foods simply *slows* growth of any microorganisms present, and that very few cells, if any, die in this condition. Any organisms present in or on the food readily begin to grow when the food is warmed to any extent. The farther below room temperature the temperature of food (i.e., if the food is frozen), the more the growth of microorganisms is inhibited and the longer it can be maintained without spoilage. Like cooling, freezing does not kill most bacterial cells and when the food is thawed, the microorganisms present will begin to grow and rapid decomposition of foods can result. Thus, when cooled or frozen foods are to be cooked, they should not be allowed to warm for long periods of time before cooking begins. Preferably, even with frozen foods, from the standpoint

Table 3.1.   Temperature Relations to Bacterial Life

°F	°C	Temperature Effects on Organisms
250	121	Wet steam temperature at 15 lb. pressure kills all forms including spores in 15 to 20 min.
240	115.5	Wet steam temperature at 10 lb. pressure kills all forms including spores in 30 to 40 min.
230	110	Wet steam temperature at 6 lb. pressure kills all forms including spores in 60 to 80 min.
220	104.5	Wet steam temperature at 2 lb. pressure
212	100	Boiling temperature of pure water at sea level. Kills in vegetative stage quickly but not spores except after long exposure.
200	93.3	Growing cells of bacteria, yeasts, and molds are
190	87.8	usually killed (180°–100° F).
180	82.2	Thermophilic organisms grow in this range (150°–
170	76.7	180° F).
170	76.6	Pasteurization of milk in 30 min. kills all the
160	71.1	important pathogenic bacteria as far as humans
150	65.6	are concerned except the spore-forming pathogens (140°–170° F).
100	37.8	Active growing range for most bacteria, yeast, and
98.6	37.0	molds (60°–100° F).
60	15.6	Growth retarded for most organisms (60°–50° F).
50	10.0	Optimum growth of psychrophilic organisms (50°–40° F).
32	0	Freezing. Usually the growth of all organisms is stopped.
0	−17.8	Bacteria preserved in a latent state.
−420	−250	Many species of bacteria are not killed by the temperature of liquid hydrogen.

Source: Weiser *et al.* (1971).

of preventing spoilage and particularly if the food is a large mass such as a ham or turkey, the cooking process should begin before thawing is entirely completed.

Pickling is a means of food preservation that has long been used. In effect, pickling accomplishes much the same thing as does drying. Pickling increases the concentration of some chemicals in foodstuffs and preserves in this manner. By pickling, salt content of food is increased following exposure to brine for a period of time; then the acid exposure (acetic acid or vinegar) adds additional preservation. Pickled foods will last for long periods without further preservative measures; however, if left exposed to air, some resistant microorganisms will grow and spoil the food.

Smoking is a preservative process used for meats, fish, and other foods. Smoking actually has double preservative action. It serves to dehydrate, and also, there is a germicidal action of chemicals that are present as a part of the smoke. The chemical preservation penetrates only a very short distance into the food. However, if the internal portion of the food was initially free from contamination, this has the effect of

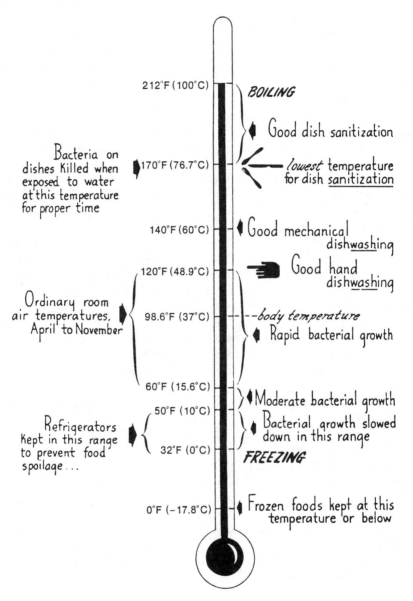

Modified from Weiser et al.. 1971

Fig. 3.1. Temperature and food sanitation.

sealing the food and it can now be maintained for considerable periods of time. If, however, the meat or fish was contaminated before smoking, and contaminated deep within the tissues, then this process may be dangerous because it may simply maintain the contaminants already present and in some instances allow later growth which results in food poisoning of certain types.

The major method of preservation for foods that are to be stored for long periods remains the canning process. Canned foods may be preserved in either small or large quantities, and until relatively recent years the home was the major source of canned foods. More recently, canning has become a major industrial process used worldwide. Canned foods are exposed to heat to destroy any organisms present, and then are sealed in order that additional contamination cannot occur. With the more modern canning processes, foods can be sealed in a vacuum chamber so that most of the air (oxygen) is removed from the cans, thus preventing staling as well as spoilage. Foods of different chemical content require different temperatures in the canning process, dependent upon the potential contamination present and the type of contamination most likely to cause spoilage.

Low acid foods generally must be processed and canned at temperatures well above the temperature of boiling. This is done by use of steam pressure cookers (autoclaves) and sterility can be accomplished only by the use of these above-boiling temperatures for the proper length of time. Some foods are damaged in texture, palatability, aroma, or even nutritional value by this process and are not satisfactorily preserved by canning. In canning, it is necessary that both vegetative cells and spores of microorganisms be killed in order that spores cannot later germinate and produce vegetative cells, thereby resulting in spoilage or in food poisoning. Some foods—those high in sugar, salt, or acid content—do not require such extreme temperatures because they have a "built-in" preservative in the form of the sugar, salt, or acid.

In a properly operated commercial canning plant, food preservation can be controlled to a point where there is essentially no danger of spoilage and little or no danger of food poisoning following consumption of the foods. On the other hand, home canning operations can present problems unless great care is taken to control prior contamination, and temperature and time used for the canning process. In home canning, as in commercial operations, accurate measure of temperatures *must* be made to assure the degree of preservation desired in that particular food. This is especially true in canning foods that are low in acid, salt, or sugar, in which spores may survive when the correct temperatures are not attained for a long enough period. In some canned foods there can be sufficient growth of some microorganisms to produce serious food poisoning although there may not be obvious spoilage. In others, extensive growth of microorganisms results in spoilage with decomposition and putrefactive odors that should prevent the food from being eaten. A few spores of some anaerobic bacteria may germinate in some canned foods, producing vegetative cells that release toxins very dangerous to the human. In the presence of such organisms, the food may not be obviously spoiled and there will be no apparent reason to discard it. Because of this danger, it is essential that all foods, particularly those canned in the home, be of good grade, in good condition, and that the canning process be carried out with good controls and efficiency.

The preservation of food is ever more important to the well-being of mankind, even though with the increased efficiency of transportation, fresh foods are more frequently used year round, in all parts of the world, rather than being used just only seasonally. Food protection is important, however, in handling and shipping foods, both from the standpoint of sanitation and of preservation. Fresh foods must be cleaned thoroughly; particularly vegetables and fruits. Such cleaning helps to remove environmental contamination. However, cleaning processes carried out in the home can serve as the source of contamination with microorganisms from the cook or food handler. In this way, infectious disease may be transmitted from food handler to the consumer, or the food may serve as a culture medium for some microorganisms that will produce toxins dangerous to humans.

MICROBIALLY CAUSED FOODBORNE DISEASES

Microbially caused foodborne diseases may be considered to be of four types. Two of these may correctly be termed food poisoning. In one type of food poisoning, growing microorganisms produce soluble, poisonous substances that are secreted into the food and these toxins produce disease in humans when the food is consumed. In this food poisoning, it is not necessary that the microorganism itself be present when the food is eaten for damage to be produced. In most cases the organism is present, but it may not survive passage through the digestive tract and it is not necessary that it continue to grow. Toxin is produced during the growth of the microorganism, before the food is eaten.

In the second type of food poisoning the organism is ingested in the food, and after reaching the digestive tract, produces the enterotoxin that causes the disease. This is still a food poisoning because it is the toxin that causes the disease, but in this case the toxin is produced at a different time in a different place. Although the organism is ingested and grows and reproduces in the intestinal tract, the disease is not produced because of invasion of body tissues by the microorganism, and therefore technically is not an infection because tissues are not invaded.

In the third type of foodborne disease caused by microorganisms, the organism itself is transmitted to the digestive tract of man. Here it grows, reproduces, and to some extent invades the body tissues to produce disease. This is an actual infection rather than the result of toxic materials produced by the organisms. Even in this situation, however, strict definition of *food poisoning* or *infection* may be confusing because some of the organisms producing this type of disease may do so in part because of the production in the intestinal tract, or nearby tissues, of toxic materials. In these cases, although the toxin is produced, it is produced only after the growth of the organism and invasion of the tissues. This type of disease requires drug treatment to remove the microorganisms rather than simple treatment of toxic symptoms.

In any of the preceding cases, the major source of disease begins in the intestinal tract, and it is here that either the toxin is absorbed to do its damage, or that the microorganisms grow, reproduce, invade, and begin to cause distress to the human; a condition called gastroenteritis, or enteritis. Some other types of infections are transmitted by way of foods and pass through the digestive tract to other organs that are invaded, and localized or generalized infections are produced. Although these infections do not actually produce disease in the gastrointestinal tract, and are not properly termed *enteritis*, they most certainly must be considered as foodborne diseases. This type of foodborne transmission of infectious disease has come to be recognized with much more frequency than in the past, and provides additional evidence that we must critically assess all stages in food handling in order to assure food safety.

A fourth type of foodborne disease caused by microorganisms may be represented by any infectious disease. In this case, food contaminated by some pathogenic microorganism is ingested, and an infection is begun by the microorganism so carried. An example of this type of foodborne disease may be milkborne tuberculosis, or milkborne diphtheria. The organism transmitted, while definitely causing disease, does not result in symptoms involving the gastrointestinal tract, and may not be easily recognized as having been foodborne.

EXAMPLES OF THE TWO TYPES OF FOOD POISONING

One of the most common, if not the most common, types of food poisoning in the United States and in most European countries is staphylococcal food poisoning (see Table 3.2). This disease is an intoxication rather than an infection. Various foods, particularly those of a creamy nature containing such ingredients as mayonnaise, salad dressing, or cream fillings, provide excellent nutritive environments for the growth of many microorganisms. The food is actually a very good, rich culture medium. Staphylococci live normally on the human skin, and in this location frequently produce very minor, well localized skin infections such as pimples or boils. In the preparation of foods it is quite easy for such organisms to be transferred to the food by way of the cook's hands. If this occurs, for example, in the preparation of potato salad, the transfer of the organisms comes after the food has been cooked. Most often the potato salad is left, at the best, at refrigerator temperature, and quite possibly at room temperature for some time—even several hours, before it is consumed. This provides ample time for the growth of the staphylococci in the food before it is eaten. The number of cells present before growth occurs does not have to be large. There will be no apparent spoilage of the food; that is, no change in flavor or aroma. In the process of growth, the staphylococci produce a potent, soluble exotoxin that is secreted into the food and causes the disease. Within two to four hours after such toxic food is ingested, nausea, abdominal cramps, and vomiting may begin suddenly and may be rather violent. These symptoms may be followed shortly by diarrhea or, in more severe cases, prostration. Staphylococcal food poisoning has been called "church picnic" food poisoning because of the common practice of taking creamy foods to picnics with the resultant exposure of the foods to warm temperatures for several hours, thereby providing incubation and permitting production of the toxin. Staphylococcal food poisoning is not a serious disease from the standpoint of causing permanent damage or death. It is an acute disease, producing extreme discomfort in the patient for a matter of hours, or perhaps one or two days and is followed by spontaneous recovery. Because the disease is an intoxication rather than an infection, antibiotics are not helpful in treatment, but rather, symptoms are treated to ease the distress of the patient. Staphylococcal food poisoning can be prevented only by taking precautions to prevent the entrance of the organism into the food, and by proper refrigeration to prevent the growth of the organisms that enter inadvertently. This type of food poisoning is also sometimes called ptomaine poisoning; however, this is a misnomer with the term having come into rather common usage in reference to any acute, violent gastrointestinal upset. Other than potato salad, foods that are common vehicles of staphylococcal food poisoning are cream-filled pastries, creamed seafood dishes, meats, gravies, green salads with creamy dressings, and in general those foods that contain rich mixtures of oils, proteins, and carbohydrates and that do not contain high concentrations of salts, sugars, or acids. Once the toxin has been produced by the organism in the food,

Table 3.2.   United States and Territories Foodborne Infection Outbreaks In 1982

Etiology	# Outbreaks	# Cases	Vehicles	Locations
Bacillus cereus	8	200	3 Fried rice 1 Steamed rice 1 Ice cream 1 Sandwich 1 Baked fish	3 Restaurants 2 Schools 1 Camp 1 Hospital 1 Fire hall
Brucella sp.	1	3	Unknown	Home
Campylobacter jejuni	2	31	Milk	Home Unknown

Clostridium botulinum	21	30	16 Home-canned
			    vegetables
			2 Commercially
			    canned
			    vegetables
			1 Beef pot pie
			1 Fermented fish
			    eggs
			1 Unknown
				Homes

Clostridium perfringens	22	1,189	11 meat dishes 4 Poultry dishes 7 Unknown	4 Prison 5 Parties 1 School 9 Restaurants 1 Hospital 2 Meeting halls
Escherichia coli	2	47	Ground beef	Restaurants
Salmonella sp.	66	2,061	29 Unknown 8 Poultry 5 Meat dishes 1 Egg dish 7 Milk products 2 Salads 1 Vegetable 2 Seafoods	13 Homes 18 Restaurants 3 Picnics 4 Hospitals 3 Nursing homes 4 Churches 2 Statewide 6 Schools 3 Camps
Shigella sp.	4	116	1 Tuna salad 1 Salad bar 2 Unknown	2 Restaurants 1 Home 1 Church
Staphylococcus aureus	28	669	10 Meat dishes 8 Poultry dishes 2 Creamy foods 2 Potato salads 3 Sandwiches 1 Vegetable 2 Unknown	5 Homes 2 Nursing homes 6 Restaurants 2 Hospitals 8 Picnics 2 Church 1 Fire hall 1 Fair
Streptococcus sp.	1	34	Onion & clam dip	Party

Vibrio sp.	5	938*	1 *Uncooked
			    shellfish
			2 Raw clams
			1 Steamed clams
			1 Shrimp
				*3 Homes
				    Clam bake
				    Picnic

Table 3.2. (*continued*)

Etiology	# Outbreaks	# Cases	Vehicles	Locations
Yersinia enterocolitica	2	198	Bean sprouts Pasteurized milk	Brownie meet Home
Trichinella spiralis	1	4	Smoked wild pork	Home
Viral Hepatitis A	19	325	5 Shellfish 1 Tuna fish 1 Salad & ice 1 Cake 10 Unknown 1 Ice cream	5 Homes 6 Restaurants 1 Hospital 2 Picnics 2 Armed forces 2 Camp 1 Party

Vegetables involved in *Cl. botulinum* outbreaks were 5: beans; 2: peppers; swiss chard; 3: asparagus, zucchini, eggplant, relish, tomatoes, and mushrooms. Salmon was involved in 1 outbreak.
* 892 Cases were in Guam in one outbreak.
Source: Foodborne Disease Outbreaks Surveillance. 1982. U.S.D.H.H.S. Public Health Service.

there is no way to remove it. Because the toxin is heat resistant, it would not be destroyed by additional cooking, even if the food were cooked again, which generally is not done.

Another type of food poisoning that may sometimes resemble staphylococcal food poisoning is that caused by *Bacillus cereus*. Outbreaks of this disease are most often associated with consumption of foods such as rice, potatoes, and vegetables, with meat dishes occasionally incriminated as the vehicle. An enterotoxin, closely associated with the bacterial cell, is produced as the organism multiplies very rapidly. Rapid multiplication and toxin production may occur in the food prior to ingestion, or it may occur in the intestinal tract following ingestion. This is the reason for the extended incubation period in some instances. There are two different enterotoxins produced by these organisms, with one likely to cause vomiting within fifteen minutes to several hours after ingestion, and the other more likely to cause diarrhea after a more extended incubation period. If foods are contaminated by vegetative cells, these cells, exposed to optimum temperatures, will undergo rapid growth and produce the enterotoxin prior to ingestion. If foods are contaminated by spores, then the spores are more resistant to the extremes of the digestive tract, and may well survive passage through the acid conditions of the stomach and the alkaline conditions of the small intestine—then vegetate and grow rapidly to produce toxin. Symptoms of this type of food poisoning seldom last more than twelve to twenty-four hours and recovery is complete.

A different food poisoning is that produced by an anaerobic bacterium, *Clostridium perfringens*. This organism lives normally as a part of the flora in the intestinal tract of humans and other animals, and does not produce disease. If it is accidentally inoculated into a wound in the deep tissues of the body, it produces an infection by reproduction and growth at the local site where it produces a soluble exotoxin. In these deep tissues the infection produced is called gas gangrene. If this bacterium happens to contaminate certain foods and is able to grow and reproduce, when the food is ingested it will cause much the same sort of symptoms as those produced by staphylococcal toxin. The large number of vegetative cells in the food, upon reaching the intestine, actively grow and produce soluble enterotoxins that cause the illness within nine to fifteen hours after eating. The acute symptoms appear rapidly with nausea, headache, abdominal pain, and diarrhea, and most commonly recovery is complete within twenty-four to forty-eight hours Like staphylococcal food poisoning, although the symptoms may be severe the disease is rarely fatal in healthy persons. Rare severe and fatal disease has occurred in some countries.

Another type of food poisoning occurs in the human as the result of growth of an anaerobic bacterium, *Clostridium botulinum*. Unlike the previously described food poisonings, botulism is a highly fatal disease. It is truly a food poisoning and results when foods are consumed in which the bacteria have been able to grow under anaerobic conditions. Most often, these conditions are likely to occur in nonacid canned foods, but smoked and dried foods sealed in airtight containers have also been incriminated. In preservation of foods, when mistakes are made that allow the survival of a few spores of *C. botulinum*, although all vegetative cells are killed, these spores may later germinate forming new vegetative cells which grow and produce the soluble exotoxin. Generally, the foods are not spoiled in the sense that putrid odors or obvious decay is produced. The toxin present in the food is not easily detected and will not be observed by the consumer.

Symptoms of botulism generally appear in twelve to twenty-four hours after the food is eaten. These symptoms indicate involvement of the central nervous system and may include weakness, dizziness, headache, and hoarseness which will be followed by some paralysis. Nausea and diarrhea occasionally occur. The diagnosis can be made by cultures and toxicity tests done on remaining portions of the food that appears to be implicated. If the diagnosis is sufficiently early, a person having consumed toxic food may be treated with specific antitoxin. The disease may be prevented by using sufficient heat in food preservation to kill any spores present so that the toxin is never produced. Food containing the toxin may be made safe for consumption by heating at 100° C for fifteen minutes before serving.

In addition to botulism, which is a food poisoning, in recent years it has been shown that an additional form of botulism exists: infant botulism. In this instance the anaerobic organism grows in the intestinal tract of the infant and produces toxin in this location. The feeding of honey to the infant has been suggested as a possible cause of infant botulism, and the condition is considered to be one possible explanation in sudden infant death syndrome. In most instances, affected infants recover with some amount of supportive therapy.

In food poisonings just discussed, disease is produced as a result of ingestion of soluble poisons produced by bacteria in food, although the food does not appear to be spoiled. If food is spoiled to any extent there may be nonspecific, undesirable chemicals present that may produce nausea, vomiting, and diarrhea of varying duration, even true ptomaines. These are not specific food poisoning diseases, and when such conditions are present there is usually sufficient decomposition present that the individual should have detected it and should never have eaten the food under any circumstances.

In addition to the toxic poisonings resulting from growth of bacteria in prepared foods, a second type of food poisoning in which the organisms are ingested and produce or release the toxins within the intestinal tract of the consumer also occurs. This disease should be correctly termed food poisoning, rather than infection because there is actually no invasion of body tissues by the ingested organism. An example of this type of disease is seen in perfringens food poisoning as was described earlier, and is in contrast to actual infections.

FOODBORNE INFECTIONS USUALLY HAVING GASTROINTESTINAL SYMPTOMS

A number of foodborne infectious diseases, while having some gastrointestinal symptoms, may also progress to systemic infections, or to infections of some organ outside the gastrointestinal tract. Such infections include typhoid fever, salmonellosis, some normal flora enteric bacterial infections, some Vibrio infections, *E. coli* infections, *Campylobacter* infections, some Streptococcus infections, Listeria infections, Brucella infections, Aeromonas infections, some parasitic infections, hepatitis, and poliomyelitis. Although the symptoms and the apparent primary infection may occur outside the gastrointestinal tract, these diseases may very well be foodborne infections. Other infections discussed in the following pages may never show any gastrointestinal symptoms, yet still can be foodborne infectious diseases.

In foodborne infections, there is one situation that makes it very difficult to control the transmission of disease from one person to another. The problem is that for some types of bacterial infections, an individual may not show any symptoms of the disease, but may be carrying disease organisms somewhere in or on the body. This carrier state is usually not detected until the disease has been passed on to another person. The carrier is healthy in all respects but is serving as a vehicle for transmission of disease organisms to more susceptible individuals. In intestinal infections, the causative organisms are generally carried in the lower intestinal tract, and the lack of proper sanitation or hygienic practices on the part of the carrier permits the transfer of organisms to hands and then to food or other objects being handled. An illustration of the damage that can be done by an undiagnosed carrier is found in the case of the cook living in New York City who became known as Typhoid Mary. To all outward appearances she was perfectly healthy; however, she carried the typhoid bacillus. While serving as a cook, she was proven to have transmitted typhoid fever to approximately 100 individuals before the carrier state was detected. The carrier state in humans can occur with other organisms in addition to the typhoid bacillus. It is fairly common in the case of many *Salmonella* infections, and is also common in the case of some strains of *Staphylococcus* species. The source of most domestic typhoid fever in the United States is thought to be the undetected chronic carrier of the bacterium. In the United States, the treatment of drinking water supplies has dramatically reduced the occurrence of typhoid infections; yet in 1955 the rate of infection in the United States was still above 1.0/100,000 population, and in 1983 there were 507 cases of the disease reported. It must be assumed that most cases result from direct or indirect contact transmission from carriers, and it is most probable that many of these are the result of foodborne transmission of the organism.

To prevent transmission of foodborne microorganisms from one person to another is to control foodborne infectious diseases. Such control is largely accomplished if persons who are infected or who carry disease organisms are not allowed to handle or prepare food for themselves or for others. The difficulty in accomplishing this lies in the fact that these disease organisms are generally not detected and recognized until they have been transmitted to and caused disease in a susceptible person. To reduce transmission in this case, and in incidental contamination as well, rigid cleanliness and sanitation procedures should be followed in all food preparation, whether in the home for an individual or in a commercial eating establishment for large numbers of persons. It is necessary in all cases that handling of food by a number of persons or customers be eliminated. This is particularly essential in cafeterias or buffet type service establishments.

When organisms are transmitted by food to susceptible individuals, the type of disease produced depends upon the specific organism, and in some cases perhaps the resistance of the individual who ingests the organism. We most commonly think of foodborne diseases as those that cause some upset in the digestive system, or gastrointestinal tract. Such diseases as a group may be termed enteritis or gastroenteritis. Symptoms of such infections include vomiting, nausea, diarrhea, and general abdominal distress. Some organisms with more invasiveness will invade body tissues and also cause fever, headache, septicemia, muscular soreness, muscular pain, jaundice or liver dysfunction, and various other symptoms depending upon the organism and the site of localization in the body organs and tissues.

Many bacteria are normal inhabitants of the human intestinal tract and generally cause no disease when present. These saprophytic inhabitants of the intestinal tract are essential to the normal function of the tract, and only very rarely cause disease. These normal inhabitants are termed *enteric bacteria* and are the organisms used to indicate fecal pollution of water and foods. Many of the bacteria present and included in this group are called coliforms, and may include *Escherichia coli*. There are, however, some strains of *E. coli* that are enteropathogenic, and will frequently produce disease, particularly in young children. Others may be termed enteric bacteria because of their common association with the enteric tract in the human or other animals, although they may be pathogenic and cause disease under conditions favorable to their growth and survival. Among these bacteria are *Enterobacter, Shigella, Salmonella, Citrobacter, Arizona, Klebsiella, Proteus, Providencia, Staphylococcus, Streptococcus,* and other genera. A genus of bacteria not included in the enterobacter group because of a number of special characteristics, which does, however, cause gastroenteritis in some individuals, is *Pseudomonas*. *P. aeruginosa* is a cause of a number of other types of infections as well as enteritis. This organism is said to be, along with other species in the genus, an opportunistic agent of infection which is frequently found to be associated with intravenous, immunosuppressive, or antibiotic therapy over long periods of time. *P. aeruginosa* is also found in approximately ten percent of normal stools, and is a cause of epidemic diarrhea in infants. Although not always incriminated, it is thought by many to be at least a participant in some cases of nonspecific enteritis.

FOODBORNE DISEASE RESULTING IN GASTROENTERITIS

The ever increasing number of specific microbially caused foodborne infectious diseases is shown in Table 3.3. This list is not all-inclusive, but it does include those specific infections that are most often diagnosed and recognized. Certainly other infections could well be included in the list, and there are many instances of enteritis that go undiagnosed as to specific cause that may, in the future, be added to the list.

An infection of the intestinal tract that has occurred in epidemic

Table 3.3. Characteristics of Microbially Caused Foodborne Diseases

Disease	Symptoms	Incubation Period	Causative Organism	Foods Commonly Contaminated	Temperature Growth Range	Ingested
Staphylococcal food poisoning	Vomiting, nausea, diarrhea	2–4 hrs.	Staphylococcus aureus[1]	Creamy foods, meats, dairy & bakery products	20°C (68°F) to 40°C (104°F)	Toxin: yes; Cells: yes*
Perfringens food poisoning	Diarrhea, nausea, but little fever or vomiting	9–15 hrs.	Clostridium perfringens[1]	Reheated meat dishes	20° C (68° F) to 45° C (113° F)	Toxin: no; Cells: yes (large numbers)
Botulism	Central nervous system involved, blurred vision, sore throat, vomiting, diarrhea.	12–36 hrs.	Clostridium botulinum[2]	Canned low-acid vegetables, smoked fish	20° C (68° F) to 45° C (113° F)	Toxin: yes; Cells: yes*
Vibrio Gastroenteritis	Watery diarrhea, abdominal cramps, nausea, vomiting, fever, chills, wound infection from contaminated water. Symptoms vary depending upon species of vibrio causing infection.	12–36 hrs. depending upon species infecting and infecting dose of cells	Vibrio parahaemolyticus, V. hollisae, V. vulnificus, V. alginolyticus, V. cholerae (01), V. cholerae (Non 01), V. fluvialis, V. mimicus	Shellfish, seafoods generally, water	20° C (68° F) to 40° C (104° F)	Toxin: no; Cells: yes

Disease	Symptoms	Onset	Organism	Source	Temperature	Detection
Aeromonas hydrophila diarrhea	Watery diarrhea, abdominal discomfort, nausea, vomiting, fever	12–36 hrs. depending upon infecting dose of cells	*Aeromonas hydrophila*	Most commonly from water, not foods	15° C (59° F) to 35° C (95° F)	Toxin: no; Cells: yes
Salmonellosis	Nausea, vomiting, abdominal pain, diarrhea. In some cases may cause septicemia and high fever will be symptom.	6–48 hrs.	*Salmonella* sp.	Poultry, eggs, milk, meats, gravies	20° C (68° F) to 40° C (104° F)	Toxin: no; Cells: yes
Shigellosis	Diarrhea, sometimes fever and vomiting	12–48 hrs.	*Shigella* sp.	Poultry, eggs, milk, meat, raw vegetables	25° C (77° F) to 40° C (104° F)	Toxin: no; Cells: yes
Escherichia coli gastroenteritis	Diarrhea, sometimes fever and vomiting.	6–36 hrs.	*Escherichia coli* enteropathogenic strain	Raw vegetables, any food	15° C (59° F) to 45° C (113° F)	Toxin: no; Cells: yes
Amoebiasis	Vary greatly from none to fulminating dysentery. Episodes of diarrhea, nausea, abdominal cramps, vomiting, loss of appetite, weight loss.	4 days to 1 year	*Entamoeba histolytica*	Raw vegetables, water, any food	20° C (68° F) to 40° C (104° F)	Toxin: no; Cells or cysts: yes
Giardiasis	Acute or chronic diarrhea. Malaise, weakness, weight loss, abdominal cramps, flatulence. Organism interferes with fat absorption.	3 days to 1 week	*Giardia lamblia*	Most commonly water. Can be any food.	15° C (59° F) to 40° C (104° F)	Toxin: no; Cells or cysts: yes

Table 3.3 (continued)

Disease	Symptoms	Incubation Period	Causative Organism	Foods Commonly Contaminated	Temperature Growth Range	Ingested
Brucellosis	Fever, may be low grade. Malaise, weakness, sweats.	1–6 wks.	*Brucella abortus, Brucella suis, Brucella melitensis*	Meats, milk, milk products	25° C (77° F) to 40° C (104° F)	Toxin: no; Cells: yes
Yersinia gastroenteritis	Abdominal pain, vomiting, diarrhea, headache	24–36 hrs.	*Yersinia enterocolitica*	Milk, ice cream, water	–2° C (36° F) to 45° C (113° F)	Toxin, no; Cells; yes
Campylobacteriosis	Diarrhea, muscular pain, headaches, vomiting rare	48–120 hrs.	*Campylobacter fetus jejuni*	Milk, milk products, seafoods, water	20° C (68° F) to 40° C (104° F)	Toxin: no; Cells: yes
Streptococcal disease	Febrile syndrome and/or nausea and diarrhea	24–96 hrs.	Group A *Streptococcus* sp.	Milk, milk products, meats, creamy foods	20° C (68° F)–40° C (104° F)	Toxin: no; Cells: yes
Trichinosis	Fever, high eosinophil count, eye edema, muscle pain	3–30 days	*Trichinella spiralis*	Insufficiently cooked pork, some wild game meats improperly cooked	Relatively narrow, 30° C (86° F) to 40° C (104° F)	Toxin: no; Cells, larvae, or cysts: yes
Hepatitis	Jaundice, gastrointestinal distress, fever	Varies	Hepatitis A virus	Shellfish, any food, water	Does not grow outside human body	Toxin: no; Virus: yes
Viral gastroenteritis	Nausea, vomiting, diarrhea	12–36 hrs.	Norwalk virus, Rotavirus, Echovirus, others	Shellfish, any food, water	Does not grow outside human body	Toxin: no; Virus: yes

62

Listeriosis	Meningoence-phalitis, bacter-emia, fever	Varies, 24–96 hrs.	*Listeria monocyto-genes*	Milk, milk products	4° C (39° F) to 40° C (104° F)	Toxin: no; Cells: yes
Legionellosis	From asympto-matic to non-descript fever to acute illness with high fever, chills, malaise, diarrhea, deliri-um, pneumoni-tis	Unknown	*Legionella pneu-mophila*	None known, water	10° C (50° F) to 40° C (104° F)	Toxin: no; Cells: yes
Pork tapeworm	Abdominal dis-comfort, weight loss, muscle pain	1–3 wks.	*Taenia solium*	Pork, insufficiently cooked	20° C (68° F) to 40° C (104° F)	

[1]Rarely fatal.   [2]Up to 70% fatal.
* Although organisms may be ingested, symptoms are due to preformed exotoxin.

proportions in the past, and that has become well known as both a waterborne and a foodborne infection, is typhoid fever. Typhoid is an acute infectious disease caused by *Salmonella typhi*. In this disease, the organism enters in contaminated water or food and passes into the intestinal tract. At this site it invades the intestinal lining, enters the blood circulation system, and the result is a continuous fever, inflammation and ulceration of the intestine, enlargement of the spleen, and frequently a "rose-spot" rash on the abdominal skin. The fever and inflammation are produced as a result of the release of endotoxin from bacterial cells as they break down in the body tissues. The bacteria producing the disease may be isolated from the blood in the early stages of infection, and from the feces in later stages. The disease occurs as epidemics following floods and natural disasters where sewage contaminates water supplies, but otherwise is a sporadic infection which is generally foodborne. In the United States, epidemics of typhoid fever rarely occur because the disease has been largely controlled by the successful practice of proper waste disposal and the purification of water supplies. The major difficulty in control of this disease has been in the detection and elimination of the carrier state, as discussed previously. Typhoid fever is a dangerous disease; however, it is successfully treated with the use of antibiotics and may be successfully controlled by continued effective waste disposal, efficient water treatment, and practice of rigid sanitation procedures by infected persons and those who help care for them. Persons who have been infected should be tested periodically to determine whether they have become carriers, and if so, should be treated to cure the carrier condition. *Salmonella typhi* is shed in the feces of an infected individual, and of a carrier. Without proper hygienic practices, the organisms may contaminate the hands and then be transmitted to any food or object handled. The most common medium for transmission has in the past been untreated or contaminated water. In the history of prevention of infectious diseases, however, there are few greater success stories than the treatment of consumable water supplies. In the United States in recent years, the transmission of typhoid fever to susceptible individuals has most often been by way of contaminated milk or foods that have been contaminated by the causative organisms from some other source. The organisms causing typhoid are capable of living for relatively long periods in adverse conditions, and for even longer periods when contaminating water or foods, although not surviving quite as long when subjected to drying.

Typhoid fever has been greatly reduced in the developed countries by the processes used in treatment of water supplies, that is, filtration, sedimentation, and chlorination. Even in relatively primitive situations, it is quite easy to use either chlorine or iodine to successfully treat potentially contaminated water, and thereby to remove any danger of disease transmission by such water. Typhoid can also be controlled by immunization, although this method is both difficult and costly to apply on a broad scale as would be needed for large populations and is therefore recommended in situations such as following natural disasters, in the case of hospital personnel who must care for the infected, and for individuals in the armed services who may be exposed to infection in various geographical locations without otherwise being able to prepare in advance.

Although typhoid fever is an infection caused by a species of *Salmonella*, there are other infections caused by other species of this genus that are commonly foodborne and that must be considered even more frequently than typhoid in food sanitation. These infections, termed Salmonellosis, are more numerous than typhoid, and are likely to come from more sources than is typhoid. For example, in 1983, there were only 507 cases of typhoid in the United States, but there were 44,250 cases of other *Salmonella* infections. Salmonellosis is commonly called food poisoning, but is actually an infection. The organism invades through the intestinal mucosa and may spread throughout the body tissues by way of the blood circulatory system. An infection with *Salmonella* sp. organisms most commonly is acute, involving nausea, vomiting, and diarrhea, and may frequently develop into an enteric fever similar to typhoid. The onset is usually sudden, with fairly severe abdominal pain within one to three days after ingesting the contaminated food or water. In Salmonellosis, there is almost always some fever, and the diarrhea and fever may persist for several days. Because the organisms grow in the intestinal tract, they are shed in the feces, and with poor hygienic and sanitation practices are spread to other individuals, to water, to food, or to various other objects.

Salmonellosis occurs worldwide and may be caused by any one of a large number of species, although perhaps a half dozen species cause the majority of infections in this country. The infection may occur in isolated cases, or in large numbers of cases in outbreaks in hospitals, nursing homes, restaurants, or schools. Except when large outbreaks occur, the initial source of the infection is frequently missed; therefore, if it was because of faulty sanitation, the error may be repeated, and additional cases of infection may occur from the same source.

It is difficult to eliminate *Salmonella* infections because there is a reservoir of the organisms in domestic and wild animals, including such pets as turtles and chickens. Salmonella infections, like typhoid, are capable of producing at least a temporary carrier condition in humans with mild or no symptoms. The transmission of various types of *Salmonella* sp. organisms in food products to the human has increased with increased handling and commercial processing of foods, particularly those processed in large volumes, such as eggs, milk, and milk products. In 1984 there was a large-scale outbreak of salmonellosis with hundreds of cases of the disease reported as being due to contaminated milk in a commercial dairy operation. In this case, once the entire dairy operation was contaminated it became even more difficult to find the exact site and mechanism of contamination, and therefore became more difficult to correct the situation and remove the danger to consumers of the products.

Salmonellae are found commonly to contaminate meat and meat products, especially poultry, because the organisms inhabit the intestinal tracts of the live animals before slaughter. Foods most commonly suspect as potential vehicles for Salmonellae are meat pies, poultry pies, sausages, lightly cooked foods containing reconstituted eggs, powdered milk, and occasionally various drug preparations—particularly those that include animal products or tissues. These foods and drugs may have been contaminated in processing by organisms from the live animal, or they may have been contaminated by the persons involved in handling or processing, or they may have been contaminated by equipment in the food plant. It is not uncommon for a carrier, an infected person, or a contaminated product to contaminate a piece of equipment and for this contamination to be passed along to uncontaminated foods being processed later. Salmonellae are able to survive for considerable periods of time, and the conditions in the processing plant or kitchen can conceivably be favorable for the growth of bacterial cells. They may therefore be incubated on such equipment and may continue to contaminate any foods processed for a long period of time unless the proper sanitary cleaning procedures are instituted. Various governmental agencies attempt to control contamination of commercially processed foods, particularly those handled in bulk, as mentioned earlier. Such products are often spot-checked in the hope of eliminating sources of potential infections. In home kitchens it is necessary that persons involved in food preparation provide thorough cooking of all foodstuffs derived from animal sources and prevent recontamination after cooking is completed. If contaminated food is brought in and handled in the kitchen, it may pass contamination on to the cook or to various objects and equipment in the kitchen during handling, so that only those organisms remaining in the food itself will be destroyed during cooking. Meanwhile, during the handling and preparation for cooking, the organisms have been spread to other parts of the kitchen, or to the cook. Personal hygiene on the part of the cook and all food handlers is essential to help control such occurrences. Only by rigid hygienic and sanitation practices, and periodic checks of food handlers to eliminate possible carrier states, can this type of infection be brought under control. With use of these practices, the number of cases of salmonellosis can be reduced rather than being increased as has happened in this country since 1977.

Another type of gastroenteritis transmitted by food, water, or direct contact with infected persons or contaminated objects is shigellosis or bacillary dysentery. This is an acute infection of the intestinal tract characterized by diarrhea, which may or may not be accompanied by fever and vomiting. In severe infections, blood, mucous, and pus may be passed from the intestinal tract in the feces. The diarrhea is severe enough that one of the major considerations of the disease is the drastic loss of water and electrolytes from the body with resulting dehydration. The organisms have very little invasiveness and most commonly, at least in the United States, the disease is self-limited and rather mild with a very low fatality rate. In young children, however, the infection may be very severe and even rapidly fatal due to excessive dehydration. Bacterial dysentery is caused by *Shigella* sp. that are spread by either direct contact or by contaminated foods or water. The contamination of foods may be by persons harboring the organism, or by flies or other insects. The disease occurs in all parts of the world and appears to be more severe in tropical or subtropical climates. There are fairly frequent outbreaks in institutions, hospitals, military reservations, or in any location where crowded living arrangements exist with a resulting reduction in efficiency of sanitation. The reservoir for this infection appears to be humans; however, some species of *Shigella* do infect domestic animals, particularly poultry. The disease can be passed from individual to individual any time during infection and usually for a few days after the most acute symptoms have subsided. Bacterial dysentery may be controlled with efficient sanitary measures applied to all cases of the disease, and by proper sanitation in food preparation and handling. It is particularly urgent that good sanitation practices be stringently applied in any situation of crowded conditions involving large numbers of people. The carrier is not common in shigellosis; however, cases existing with very mild symptoms may not be recognized and such an individual may serve the same purposes as a carrier in being extremely dangerous to the public health if employed in processing, preparation, or serving of foods. In control of *Shigella* infections, particular attention should be paid to the provision and use of handwashing facilities for food handlers, and to the protection of foods against contamination by flies and other insects. With such simple precautions as these, the spread of bacterial dysentery can be greatly reduced from the 19,719 cases reported in this country in 1983. When any case appears, the causative organism should be identified and reported, and measures taken to prevent the occurrence of large-scale outbreaks.

Since the early 1970s in the United States, more attention has been paid to the infections caused by a group of organisms that were once considered to pose little or no problem. These are the *Vibrio* sp., including *V. cholerae*, which we did not diagnose between 1911 and 1973 in the United States. The first vibrio to be considered in the United States as a cause of bacterial enteritis in recent times was *V. parahaemolyticus*, a bacterium that was previously recognized as a cause of foodborne infectious disease in countries where seafood was consumed raw. Although for many years the organism had been isolated in coastal waters off North America, it was not considered to pose any problem because it was not customary to consume uncooked fish. When the problem was investigated, however, it was found that the organism can also be transmitted by inadequately cooked fish and shellfish that come from coastal waters where the bacterium lives. *V. parahaemolyticus* infection is characterized by diarrhea and abdominal cramps and in some cases nausea, vomiting, fever, and headache. The symptoms begin most often twelve to twenty-four hours after consumption of contaminated seafood. Recovery is generally within one week, although rarely there is a generalized spread of the infection throughout the body and death may occur. In studies in the past fifteen years, it is more and more commonly recognized that this and other *Vibrio* species of bacteria may be natural inhabitants of some coastal and brackish waters where seafoods are caught, and that their presence may not indicate fecal contamination of the harvest areas.

It was actually a cause for some pride that we could, until 1973, state in all truthfulness that *V. cholerae* infection had not been diagnosed in the United States since 1911, other than a few laboratory acquired cases. In that year, one case of cholera was diagnosed in Port Lavaca, Texas for which no source was ever found. That case, followed by additional diagnoses of *V. cholerae* infection in the United States from 1978 through 1986, have forced the revision of the consideration that *V. cholerae* infections do not occur in this country. Large outbreaks of *V. cholerae* 01 infections have not been observed in the United States, and in several of the outbreaks that have occurred, it has been found that the organism was foodborne, while in others it was waterborne. Cholera is an acute enteritis. Following ingestion and passage through the stomach, the organisms grow on the epithelial cells lining the intestinal mucosa without invasion of adjacent tissues. *V. cholerae* prefer an alkaline environment, and any condition in the patient that reduces the acidity, or increases the alkalinity of the stomach contents may enhance the opportunity for infection when the organisms are ingested. The bacteria produce a potent enterotoxin that causes enormous secretion of water into the intestinal tract through the intestinal mucosa, and it is this that results in the watery diarrhea characteristic of the disease. Although there is no invasion of body tissues, the organism does grow prolifically in the intestine, and the disease is severe because of the drastic loss of fluids and electrolytes from the intestinal tract. Untreated, death rates may be as high as fifty percent, but with proper treatment the fatality is generally below one percent. Asymptomatic and mild cases may be observed, and these are a major problem in control of the disease, particularly in large-scale outbreaks which occur in various parts of the world.

When cholera cases began to appear in the United States in the 1970s, it brought a new awareness of the potential problems associated with this disease, and stimulated more intensive studies of these and related organisms. In these studies, it has become generally accepted that *V. cholerae* non-01 organisms are autochthonous (that is, natural) inhabitants of many coastal and brackish waters in Europe, Asia, and the United States. These are the organisms previously referred to as the non-cholera vibrios or the non-agglutinable vibrios, but which are now included as members in the species *V. cholerae*. These organisms are recognized as capable of causing outbreaks of gastroenteritis in humans. Different clinical syndromes have been described in these illnesses. The symptoms may include diarrhea or not, vomiting or not, low-grade fever or not, abdominal cramps, and general discomfort. Diagnosis is made by isolation of the organisms from the patient's feces. Unlike infection by *V. cholerae* 01 strains, there have been reports of isolation of these organisms from human cases in sources other than the feces, indicating that some strains of these bacteria do possess some invasiveness and generalized infection can occur. Some strains of non-01 vibrios have been shown to be toxigenic, while others have shown no toxin production. Non-01 infection has most often been associated with consumption of contaminated, improperly cooked seafoods, or of raw shellfish. Incubation periods are usually twenty-four to forty-eight hours, and recovery is generally complete within two to five days with supportive treatment to prevent dehydration. The carrier state for these organisms has been demonstrated in the human. The infection in the adult is usually self-limited with supportive treatment and replacement of fluids and electrolytes, but recovery may be speeded by treatment with antibiotics.

Other species of *Vibrio* including *V. hollisae, V. alginolyticus, V. fluvialis,* and *V. mimicus* are also recognized as the cause of human gastroenteritis in some instances. These organisms also are natural inhabitants of the coastal and brackish waters of many areas, and may contaminate improperly cooked seafoods, and certainly may be concentrated by filter-feeding shellfish, which are frequently consumed raw. Symptoms of these enteric infections are generally watery diarrhea, abdominal cramping or discomfort, low-grade fever if any, and sometimes nausea and vomiting. These symptoms generally appear within twenty-four to forty-eight hours and with supportive treatment, recovery is usually complete within four to five days. Recovery may be speeded by use of antibiotics to hasten the removal of the organisms, thereby reducing the need for fluid and electrolyte replacements.

In addition to the well-recognized forms of bacterial gastroenteritis, there are occasional, small-scale outbreaks of intestinal disease due to bacteria more often considered normal intestinal inhabitants. One such bacterium that causes disease in humans is *Escherichia coli*. *E. coli* is a part of the normal flora of the intestinal tract of the human, and is found there and in rather large numbers associated with the skin and the human habitation in general. Most strains of *E. coli* are considered to be non-pathogenic and will not cause infection in the absence of some other underlying disease. There are, however, numerous strains of *E. coli* that are termed *enteropathogenic* and that do produce gastroenteritis in humans, particularly in children. The infection produced resembles in some respects the disease produced by some strains of *Salmonella*. The symptoms may include nausea, vomiting, diarrhea, fever, and loss of body fluids which will result in dehydration. In adults, this infection is generally less severe than in children, and is most likely encountered when traveling. The adult may well serve as a carrier and source of infection for susceptible children. When a child suffers from enteritis, and other, better known pathogens are not found, the possibility of *E. coli* gastroenteritis should be considered. This is particularly critical in nurseries, hospitals, nursery schools, and kindergartens where the organism may be spread both by direct contact and by contaminated food. To prevent foodborne infection due to this organism, the usual sanitary precautions should be observed, and should be particularly emphasized with food handlers in the institutions and groups just mentioned. Like *Salmonella* sp., and *Shigella* sp., *E. coli* are susceptible to some antibiotics and the disease can be treated successfully with these drugs. The patients, as in all diarrheal diseases, should be observed and water and electrolyte balances should be controlled carefully, particularly in very young children. Besides the production of gastrointestinal infections by *E. coli*, these bacteria also produce sporadic cases of infections in other body organs. It is not uncommon to detect *E. coli* in the urinary tract. Because the organism is found as a part of the normal human flora, its control and eradication is most difficult, if not impossible, which points up the need for consistently adequate sanitary hygienic practices in all relationships between persons, regardless of occupation and regardless of proximity or closeness of relationship to other individuals.

Organisms closely related to, and previously grouped with the vibrios, have been increasingly recognized as the cause of foodborne enteritis in recent years; *Campylobacter jejuni, Campylobacter coli,* and *Campylobacter fetus.* These organisms, most frequently *C. jejuni,* cause an enteritis infection similar to *Shigella* infection. The organisms multiply in the small intestine after ingestion, invade the epithelium, and produce inflammation which may result in both red and white blood cells being released in the stools during the diarrhea. Occasionally, an enteric fever may develop, although it is likely to be mild when compared to diseases such as typhoid; but it may persist for several weeks without specific treatment. The symptoms generally include diarrhea (perhaps bloody), abdominal pain, headache, malaise, and fever. The infection may be treated successfully with several antibiotics, but most cases terminate without the necessity of antibiotic use. The carrier state has been recognized, but the importance of this state in transmission of *Campylobacter* infection is not known.

Another form of gastroenteritis in the human is that caused by *Yersinia enterocolitica* or *Y. pseudotuberculosis.* The symptoms include diarrhea, abdominal discomfort, low-grade fever, headache, vomiting, and possible abscesses and septicemia. The enteritis form of infection is most often produced by *Y. enterocolitica,* and the more often fatal septicemic form by *Y. pseudotuberculosis.* The mode of transmission is not completely understood, but generally it is considered to be by the anal-oral route, or through food and water contamination. There is also a good possibility for contact transmission, particularly to children from pets. The incubation period for the infection varies from two days to one week, and treatment is generally successful with a variety of antibiotics after specific diagnosis.

Many organisms live in the intestinal tract in the normally functioning state of that tract, and generally may be called a part of the normal flora. Among these are *Escherichia, Pseudomonas, Proteus, Klebsiella, Enterobacter, Serratia* sp., and perhaps others. These bacteria are frequent inhabitants of the intestinal tract and are commonly shed in human feces. In most cases there is reason to doubt that these organisms produce specific gastrointestinal disease; however, in some cases, investigators have been led to suspect that one or the other of these groups, perhaps in combination with another bacterium, may be responsible for a particular episode of intestinal upset. In any event, it is most unlikely that any of these groups of organisms will produce a large number of disease cases or even small epidemics. Disease that is produced will be of a sporadic nature and will involve only isolated cases. In spite of this, the contamination of food, water, or milk with these organisms is undesirable and may cause difficulty, even in a small number of individuals. Like the pathogenic *E. coli,* the other bacteria are occasionally observed to cause other types of infection, including infections of the urinary tract and of the skin. Like *E. coli,* these organisms are found on the skin and in the intestinal tract as normal flora; however, they may present a serious problem if allowed to infect skin wounds, abrasions, or surgical wounds. All may present serious problems when allowed to produce infections in burns on the external skin of man; particularly species of *Pseudomonas* and *Proteus.* Like *E. coli,* these organisms are frequently present in the environment, including the intestinal tract and skin, and their control or elimination is impossible. It is therefore essential that good sanitary and hygienic practices be carried out, not with the hope of elimination of the organisms, but with the hope of keeping the number of cells at a level where disease production will be unlikely. With such practices, occasional sporadic disease resulting from so-called normal flora will be reduced and the chance of large-scale epidemics caused by pathogenic organisms will also be considerably lessened.

A rare form of gastroenteritis is caused by some strains of *Aeromonas hydrophila,* a normal water and soil inhabitant. This organism is not uncommonly isolated from foods, and is a rare inhabitant of the human intestinal tract. As a pathogen, it is seen most often as a cause of septicemia following wound infections in persons who have been in contact with water, its natural habitat. The organism has also been seen as a cause of nosocomial infections in persons with lowered resistance in hospital environments. The bacterium is sensitive to some antibiotics, and successful treatment depends upon early and correct diagnosis.

Beta hemolytic streptococci occasionally contaminate milk and other foods and the result is foodborne infection, although this infection does not always cause gastroenteritis symptoms. The transmission of these pathogenic organisms by foods may result in the production of any of the usual infections caused by streptococci, that is, strep throat, scarlet fever, erysipelas, and perhaps the most common, pharyngitis. When the latter type of foodborne streptococcal infection occurs, generally a high percentage of consumers are infected, and the incubation period is fairly short following exposure (1–4 days). Streptococci may contaminate milk, because it is sometimes a cause of mastitis in cattle, or it may be inoculated into milk or other foods by persons with a superficial skin infection, or an infection of the upper respiratory tract.

The most frequent vehicles of streptococcal transmission have been found to be egg, potato, or meat salads as opposed to milk because adequate pasteurization of dairy products has been commonly practiced. The consumption of raw milk or raw milk products, however, remains a danger from the standpoint of transmission of streptococcal or other bacterial infections.

FOODBORNE INFECTIONS HAVING OTHER THAN GASTROENTERITIS SYMPTOMS

Erysipelas is an acute infection characterized by chills and fever. An initial lesion or sore may appear as a small bright red spot at the site of infection. The skin becomes red and swollen, and the face or legs, most commonly, covered with small lesions. The infection in an individual may have come from contact with streptococci from his or her own respiratory tract, or it may have come from contact with another person who was a carrier. Contact with the clothing, personal linen, or personal articles of another individual may spread the organisms. The bacteria, generally, are susceptible to antibiotics and the infection can be treated fairly easily. However, the disease itself may be prevented by good personal hygiene, including avoidance of contact with the personal articles of other individuals, and thorough cleaning of the hands or body following exposure to individuals who may be harboring the disease or the organisms.

Impetigo contagiosum, or impetigo, is sometimes described by physicians as a purulent dermatitis. This description implies that lesions appear in the skin which may become crusted and ulcerated, generally containing some pus. The same type of infection may be caused by another bacterium, the staphylococci, and, like erysipelas, this is transmitted by direct contact with a contaminated person or article. Children are more generally susceptible to this infection, and epidemics are fairly easily started in schools, camps, and among athletic teams unless transmission is controlled by good sanitary and hygienic practices. The incubation period for this infection is fairly short, perhaps no more than two days. Although the disease is rarely serious, it may disfigure by scarring the skin. Treatment with ointments or with systemic antibiotics is usually effective; however, avoidance of the infection, again, involves nothing more strenuous than simple cleanliness and good sanitary practice.

Streptococci cause as large a variety of infectious diseases, as well as a large variety of post-infection complications, as any other single group of bacteria. One of the more common diseases produced is the sore throat or strep throat. Streptococci producing this disease, like the other organisms producing pneumonia, are easily spread by a cough or a sneeze from an infected individual, or by a carrier who has no apparent infection. Streptococci live in the throat and upper respiratory tract of humans and in many individuals do not produce apparent disease. Related organisms are also present in cattle and are commonly present in milk. If the organisms are truly cattle strains rather than human strains, however, the danger is not great. Recent studies have indicated that the human is most likely to pass human strains to pets and domestic animals, and then to receive these human strains back from the animals. Most commonly, in the case of milk or other food products, the bacteria are transmitted from humans to the food and back to susceptible human.

The streptococci are quite susceptible to antibiotics. Penicillin and many other antibiotics are used to successfully treat these infections. Again, the simple practice of covering mouth and nose when coughing or sneezing greatly reduces the spread of these bacteria.

Some streptococcus infections, such as strep throat, also give rise to complications such as rheumatic fever, an involvement of the heart tissue; nephritis, an involvement of the kidneys; or osteomyelitis, an involvement of the bone. These complications may be dangerous, and it may be worthwhile to put individuals who are highly susceptible to streptococci, that is, those who have frequent sore throats or other strep infections, on routine preventive antibiotic therapy. Improvement in hygienic practices, which would limit the spread of the organisms, is also extremely helpful to these people by reducing the chance of contact, and therefore the threat of infection.

Streptococci may produce a serious infection called scarlet fever. Some strains of the organisms produce a soluble toxin that is released into the circulating blood and spread throughout the body tissues, thereby producing the symptoms of scarlet fever. The disease does not occur in infants under six months old, or in adults over fifty, and certainly not all persons exposed to the organism will come down with the disease. The differences in susceptibility are not completely understood. Scarlet fever is no longer a common disease in this country, perhaps because of the use of antibiotics. However, if the disease occurs, isolation of the patient is advisable for fairly long periods. The symptoms of scarlet fever begin with sore throat, vomiting, fever, and finally a rash that extends to all parts of the body surface. The rash usually subsides within a few days. Antibiotics are recommended to kill the bacteria. Although the rash may be helpful to the physician in diagnosing the disease, it may also be confusing because it is similar to that present in measles and in some allergic reactions. However, a simple skin test injection serves to distinguish the rash of scarlet fever from that of other disease conditions. Scarlet fever, like sore throat caused by streptococci, may lead to the same serious complications mentioned earlier. Treatment should be supervised by a physician, with the rigid routine prescribed being followed, and reinforced by careful handling of personal articles and maintenance of sanitation in the environment where scarlet fever has been present.

A type of *Vibrio* infection that is very different from those previously described is that caused by *Vibrio vulnificus*. This organism also may be transmitted by contaminated seafoods and shellfish, and is likely most often transmitted by oysters, but may also be transmitted by contact with contaminated waters when there are existing skin abrasions or wounds. Whether it is caused by ingestion, or infection by way of a pre-existing wound, onset of the disease frequently resembles a septicemia. Those individuals infected through wounds develop swelling, erythema, and frequent vesicles in areas adjacent to the wounds, sometimes followed by necrosis. Generally there is a fever, and frequently chills. In some cases blood cultures will be positive for *V. vulnificus,* but the organism is more easily cultured from the wound and associated lesions. If diagnosis is made early, the infection is treatable with antibiotics, and recovery is generally within several days. In patients with wound infections there is frequently some additional underlying disease, and the successful treatment is not as certain. In some severe cases, amputation of limbs has been deemed necessary in treatment of the disease, which progresses rapidly whether in the presence of underlying disease or not. The rapid progression of the infection may be seen in the relatively short incubation period (12–24 hours) for an infectious process.

In 1985 there were multiple outbreaks of transmission of *Listeria mono-cytogenes* in milk products that included raw, or unpasteurized milk. In this case, as in the case of streptococcal infections transmitted by contaminated foods, the infection produced is not classified as a gastroenteric infection, but instead is a septicemic or meningo-encephalitic infection. The infection can be passed from the mother to the fetus during pregnancy and the infants so infected may be stillborn, or develop meningitis in the neonatal period. These outbreaks were among the few traced to contaminated food sources in this country, and serve to emphasize the importance of consuming only pasteurized milk or milk products.

Also classified as foodborne and contact transmitted are bacteria of the genus *Brucella. Brucella abortus, B. melitensis,* and *B. suis* are the organisms that cause brucellosis. The disease is also known as "Malta fever," "milk fever," or "undulent fever." This disease has been an important public health problem that has been largely brought under control in the United States by programs that have included slaughter of infected animals and vaccination of susceptible animals. The infection is most often transmitted to the human as a result of drinking raw milk, eating unpasteurized milk products, or consuming improperly cooked meat from infected animals. Pasteurization, properly carried out, and cooking destroy the organisms and prevent transmission of the disease by these animal products. In other instances the infection may be spread by contact to butchers, veterinarians, or farmers, in whom the organisms enter the human body through cuts or abrasions in the skin. The organism could also infect the food handler in the kitchen in the same manner.

An infectious disease organism that is found worldwide in freshwater sources and that causes primarily a respiratory tract problem is caused by *Legionella pneumophila*. Although there are no recorded cases of this organism being transmitted by foods, since its normal habitat is water, and because it is frequently isolated from air-conditioning systems and washing facilities in buildings, it could theoretically become foodborne. Infections are primarily of the respiratory system; however, there is usually a generalized febrile illness of varying duration, with high fever, and chills; diarrhea may also be associated with the symptoms. The organisms are susceptible to some antibiotics and if diagnosis is made early, treatment is fairly successful in patients not otherwise suffering from debilitating disease. The infection is not transmitted from human to human; therefore, it appears unlikely that the carrier condition will have any effect in this case.

FOODBORNE DISEASE CAUSED BY PARASITES

A different type of infectious disease is that caused by single or multicelled animal parasites. Some of these infections are of the gastrointestinal type, while others are infections of other parts of the body. Many of the parasitic infectious diseases may very well be foodborne; however, we will consider here only those that are primarily gastrointestinal in nature, or those that are solely foodborne. One example of parasitic gastroenteritis is that caused by *Entamoeba histolytica*. In 1983 6,658 cases of this infection were reported in the United States. This is up from 3,329 cases reported in 1943, and from 2,235 reported in 1973. This increase is not believed to be due to a dramatically increased incidence, but rather to the importation of cases in the influx of refugees to this country after the early 1970s.

Amoebic dysentery, the infection caused by *E. histolytica*, in some ways resembles bacterial dysentery, although it is more likely to exist for long periods of time as a subclinical or mild infection in some individuals. The infection exists primarily in the large intestine and symptoms may vary from no symptoms, to mild abdominal discomfort, to severe diarrhea with the expulsion of mucous and blood from the intestinal tract. The infection in the human may progress from the intestinal tract to produce an extensive, systemic disease resulting in abscesses of liver, lung, brain, or ulceration of the skin. Such spread is relatively rare, however, and infection by amoeba is a relatively rare cause of death. In the intestinal infection, the organism and its resistant form, cysts, are spread from the feces, and susceptible individuals are infected, either by direct contact, or by consuming contaminated water or food. Contaminated vegetables which may be served raw are a common source of infection, or people may be infected by consuming food that has been contaminated by food handlers, or by flies carrying the organism. A true carrier state apparently does not exist; however, the infection in a chronic case may be without symptoms, and this individual may serve as the source of infection for others. The disease is more common in tropical or subtropical climates. It may be controlled only by proper sanitary precautions, and perhaps the major control is and has been the protection of public water supplies against fecal contamination. Water treatment, including chlorination, and checking of food handlers for presence of the organism followed by maintenance of rigid sanitary conditions in all food preparation are essential to prevent spread of this infection. This, of course can be accomplished only with the proper health education of the general public, and under the supervision of health agencies. In the United States, the actual contamination of vegetables that may be consumed raw is relatively rare. Should we become careless in our consideration of disease, however, this always remains a possible source of infection.

A second parasitic intestinal tract infection that is common worldwide is giardiasis. This infection is caused by a protozoan, *Giardia lamblia,* and is an infection of the small intestine. Although symptoms frequently are inapparent, the disease may also be characterized by chronic diarrhea, abdominal cramps, bloating, fatigue, and weight loss. Malabsorption of fats may occur, which is responsible for the greasy appearance of the stools in some cases. The infection is most often transmitted by contaminated water, but as in other cases, if water can be fecally contaminated, foods may also be contaminated, and the foodborne route of infection cannot be ignored. Some drugs are beneficial in treatment of the infection; however, relapses can occur with any drugs currently used. The disease is best controlled by sanitary disposal of feces, and by efficient water treatment. There is no immunization for prevention of the infection.

An additional parasitic disease involving the intestinal tract that is seen worldwide is the infection by the pork tapeworm, taeniasis. This infection is caused by ingestion of either the adult or larval stage of the pork tapeworm *Taenia solium*. Symptoms of the infection are extremely variable, and in many cases, not noticeable. When present they may include loss of weight, abdominal pain, and digestive disturbances. The larval stage of the infection may become generalized; larval invasion of tissues may produce very serious disease and/or complications when it occurs in vital organs of the body. The infection may be transmitted by ingestion of insufficiently cooked pork containing infective larvae, by ingestion of contaminated food or water, or more directly by contact with an infected person. Following infection there is no development of resistance, but several drugs show some improvement in the treatment of cases.

Another parasitic disease also transmitted by improperly or insufficiently cooked pork, as well as other meats, is trichinosis. This infection is caused by ingestion of encysted *Trichinella spiralis* larvae in improperly cooked meats, which then migrate to muscles within the human body and encyst in this location. In humans it is usually a mild febrile illness; however, it may be a rapidly developing fatal disease. The symptoms include sudden swelling of the eyelids, which may be preceded by gastric disturbances such as diarrhea, and there may also be chills, weakness, and intermittent fever. Fever is usually terminated after one to three weeks, when there may be an appearance of respiratory and neurological symptoms. In the milder cases, the infections are usually self-limited; however, when treatment is required, there are new drugs that are safe and effective for use in this disease.

FOODBORNE DISEASES CAUSED BY VIRUSES

A group of infectious diseases that have not always been well understood or well diagnosed are intestinal diseases of viral origin. Several viruses may enter the human body through the mouth, and may or may not produce symptoms of intestinal disease before proceeding to localize and infect organs in other parts of the body. However, there are also several viruses that enter the intestinal system in contaminated foods or water and produce gastroenteritis without progression to other organs. The most common illustration of this group of infections is the disease commonly called the virus by the general public. This infection is actually caused by a parvo-virus or parvo-viruslike–agent, and is properly called viral gastroenteritis. The symptoms of such infection are not unlike the symptoms of other infections of the gastrointestinal tract, including vomiting, nausea, diarrhea, and in some instances, fever. The infections may be caused by several virus strains that have been identified and studied in relatively recent years, including the Norwalk virus, Rotavirus, Echovirus, and others that may differ only by being different types of these viruses. The infections appear to travel rather rapidly from one human to another; however, infections of large numbers of individuals at one time by consumption of contaminated food or water is possible. The presence of an intestinal tract disease that cannot be explained by bacterial or parasitic identification should always call for stringent sanitary measures in a household, or in any other group. A person suffering such an infection should never be involved in preparation, serving, or handling of foods or drinks. Special sterilization processes should be used for eating utensils, articles of personal clothing, or linens to prevent spread of infection between persons. This type of virus gastroenteritis is generally rather short-lived, and the only effective therapy is fluid and electrolyte replacement to avoid severe dehydration. The sudden appearance of the symptoms of nausea, vomiting, and diarrhea permit the initial spread of the infection between persons before adequate precautions are deemed necessary. The symptoms generally last only for relatively few hours, and recovery is rapid.

Viral gastroenteritis in infants and very young children is not clinically differentiated from that just described, although it is now thought to be caused most often by a reoviruslike agent. Any diarrheal disease in the very young is an extremely serious condition and must be treated as such. Great care must be observed in the maintenance of electrolyte and fluid balance in these patients. The disease is so widespread that in some studies it has appeared that most children have acquired some immunity in the first three years of life. Particularly with children, the possible direct transmission of infection between persons exists; however, sanitation practices in the home should be as stringent as possible when such infections exist.

A more serious viral infection is hepatitis. At one time this disease was divided into infectious hepatitis and serum hepatitis. More recently, however, it has been recognized that there are multiple types of hepatitis infection, with different causative agents. Infectious hepatitis is now referred to as Hepatitis A, or viral hepatitis, type A. The onset of infection is generally sudden with fever, malaise, nausea, and abdominal discomfort. The jaundice, or yellowing of skin and eyes, generally follows within a few days. The infection may vary from a relatively mild disease lasting only one or two weeks, to a more severe and disabling disease which may last for several months. Convalescence from the infection is generally prolonged. The infection may weaken an individual to the point that there is greater susceptibility to other infections and diseases. In the presence of this infection, stringent sanitary practices are essential, and an infected or convalescent person should not be involved in preparation or service of foods under any circumstances. Personal articles and particularly eating utensils used by infected individuals should be routinely sanitized. Recovery is generally complete within two to three months, and the disease is not often fatal.

Hepatitis A is caused by the HAV virus, a member of the picornavirus family. It is primarily transmitted by person-to-person contact, but as in the case of all such infections, it can also be transmitted by contaminated food, water, and articles, termed *common source epidemics*. The carrier state with the HAV present in the flood or feces has not been demonstrated. Diagnosis of the infection is frequently symptomatic, but may be made definitely by demonstration of antibodies in the blood serum. The incidence of this infection has decreased in recent years; however, it is still rather common in older children and young adults.

Hepatitis B, previously called serum hepatitis, is caused by the HBV virus, and is a major cause of chronic hepatitis, cirrhosis, and certain types of liver cancer worldwide. It has an incubation period of 45–160 days after exposure, and the onset of symptoms is generally not as sudden as in hepatitis A, although the symptoms are not greatly different from those. Close personal contact, even sexual contact, is thought to be essential for transmission of the infection, or accidental inoculation with soiled instruments or articles in medical settings can transmit the disease. It is not thought that contaminated food or water transmit the infection to uninfected individuals. A preventive vaccine is now available for use in high risk individuals.

Delta hepatitis, caused by the HDV virus, is now recognized as occurring as a co-infection, or a superinfection with hepatitis B. It generally causes an episode of acute hepatitis. The co-infection state generally is overcome; however, the superinfection frequently causes a chronic case of disease. Either type may be fulminating and fatal. Non-A, Non-B hepatitis has been observed to occur in the United States most often after parenteral drug abuse or blood transfusions. The infection is similar to hepatitis B in symptoms and in transmission; however, there does not appear to be any immune cross protection. There does appear to be an epidemic form of this type of hepatitis, but it has not been recognized in the United States or Western Europe. This form of the disease is transmitted by the fecal–oral route, and therefore food contamination may possibly lead to transmission.

Poliomyelitis is not recognized as an intestinal disease; however, there is evidence that the virus enters the body by the intestinal route, and the polio vaccine is given orally. Milk has been incriminated as a vehicle of transmission; however, other foods have not been incriminated. It is thought that personal contact, or perhaps contaminated water may be the source of some outbreaks. The fact that transmission of the disease can occur by way of the intestinal tract emphasizes that sanitary practices are necessary in all aspects of preparation of foods or beverages for human consumption when there is a possibility of presence of polio virus contamination. By intensive campaigns for immunization, the incidence of polio has been dramatically reduced in the United States. However, the continued occurrence of even a few cases presents a danger to those individuals who are not resistant to the infection.

OTHER DISEASES THAT MAY BE FOODBORNE

In the foregoing, the diseases discussed may be considered to be either primarily intestinal or foodborne diseases, although some of them are generalized infections that happen to be transmitted by contaminated foods under certain circumstances. A number of infections of the respiratory tract must be considered to be potentially transmitted by contaminated foods. Diseases of the respiratory tract occur in epidemic form in many cases, attacking large numbers of persons within a very short period of time. These infections frequently cause some symptom that makes it very easy for the causative organism to be transmitted from the infected individual, for example, the cough or sneeze. In some cases the organisms producing such disease may have the ability to infect more than one part of the body, may be able to survive well on certain foods, and may enter the body by several different routes in order to cause different types of infections.

Perhaps one reason for the frequent occurrence of airborne respiratory tract disease is the relative ease of transmission of such infections from one individual to rather large groups. The incidence of these diseases is greatest in the cooler months of the year; however, they certainly are not limited to these months by any characteristic of the organism, but rather by the ease of spread. Coughs or sneezes are instrumental in the rapid spread of large numbers of organisms carried in water droplets from the respiratory tract of an infected person to additional individuals. A sneeze that is not impeded in any manner may carry water droplets and thus many hundreds of thousands of bacterial cells, for a distance of up to fifteen meters from the source. It is obvious, therefore, that in a room with many people present, large numbers may be contaminated by one sneeze. It is also obvious that a sneeze or a cough by a person who is preparing, handling, or serving foods may contaminate foods at considerable distances. In serving foods, particularly to groups such as in cafeterias and schools, or in public eating establishments, especially where buffet-style service is utilized, the person being served, or serving, may completely spoil whatever sanitary precautions have been previously observed, by one cough or sneeze unimpeded in any fashion. Many infectious organisms can survive for fairly long periods in prepared foods and, therefore, the importance of good hygienic and sanitary practices of persons who have infections of the upper respiratory tract is quite apparent.

For many, one of the more dread diseases over the years has been tuberculosis. It is an endemic disease worldwide, and is still a leading cause of death. It does not appear to occur more frequently in one season of the year than another and because it is frequently a chronic disease, at least in the pulmonary form, it may exist in an individual for several months before it is recognized. The most common form of tuberculosis is the pulmonary infection; that is, infection of the lungs. However, tuberculosis infection may occur in any tissue of the body. In humans, the disease advances slowly, symptoms may not be apparent, and a chest X-ray quite frequently is the first indication that something is wrong. In some persons there may be vague chest discomfort, sometimes with a cough, occasionally with fatigue and weight loss, before the chest X-ray spots the disease. The chest X-ray is able to detect the infection because the organism, *Mycobacterium tuberculosis,* produces a lesion in the lungs called a tubercle. This is actually a result of body cells that move into the lungs in an attempt to wall off the bacteria. If the body cells are successful in walling off the organism, the disease becomes inactive and we say that the infection is arrested. Until this occurs, the abscesses may grow in size, and may drain and spread the bacteria to other parts of the lungs. Physicians use several methods for diagnosing tuberculosis, the best of which is to demonstrate the bacterium present in the sputum, the washings of the stomach, or other body fluids and tissues. A skin test is sometimes used (the tuberculin test); however, neither the X-ray nor the tuberculin test alone is sufficient for diagnosis.

Treatment of tuberculosis generally consists of bed rest, adequate diet, and the use of effective drugs and antibiotics. This treatment has sharply reduced the number of deaths from tuberculosis; however, it has not greatly reduced the number of tuberculosis cases. Perhaps part of the reason for this is that the early symptoms of the disease are either inapparent or so mild that many cases go undiagnosed and during this period the organisms may be spread by discharges from the upper respiratory tract, by coughs or sneezes. The tubercle bacillus is relatively resistant and may survive for considerable periods outside the human body and in foods, which helps to increase the spread of the organism. Because of these characteristics, good hygienic and sanitary practices are made even more important in the control of tuberculosis than in some other infections. Here, perhaps, the cough is a major source of spreading the organism, thus producing contamination of the environment and other individuals. The simple hygienic practice of covering the mouth helps to reduce this risk. Routine checkups by a physician, which make possible early indication that tuberculosis may be present, greatly help to limit the number of cases, if these checkups are sought by large numbers of individuals. Early diagnosis makes possible early treatment and isolation of infected individuals, which helps reduce the spread of the disease. One of the greatest difficulties in controlling this infection is the fact that in spite of our knowledge of the disease and successful treatment, there still occurs an attitude of social stigma attached to tuberculosis. Some individuals in the United States, even in this day of supposed enlightenment, consider that "nice" people simply do not have tuberculosis. Occasionally, it is even found that some people consider tuberculosis to be inherited. It is true that if one's parents are infected with tuberculosis, anyone living with them has an excellent chance of being contaminated with the organisms and of contracting the infection. In addition to the personal hygienic practices mentioned earlier, it is essential that good sanitation be practiced any time a known or even a suspected case of tuberculosis is present. Such practices would include (a) coughing into disposable materials, e.g., tissues; (b) the collection of these tissues in a covered container; (c) the sterilization of these tissues with disinfectant; and finally (d) burning—incineration of the tissues. Articles of clothing, eating utensils, personal articles, and the room where an infected individual stays, or even the house where he or she stays, should be disinfected thoroughly. The tubercle bacillus coughed up into a room or onto a floor and allowed to dry may survive for some time. In drying, it may eventually be stirred up with dust and spread. In modern, air-conditioned buildings, this dust may be blown into ventilation systems, air-conditioning systems, and may collect to be spread throughout the system that is not properly constructed or operated. It may thus contaminate a large area. Disinfection, careful personal habits, and good sanitation help to reduce the spread of contamination in any environment, and therefore reduce the potential for production of tuberculosis in a great many individuals. Only with this practice of personal hygiene and sanitation, along with the practice of frequent physical checkups to permit early diagnosis, do we have good hope of significantly reducing the number of cases of tuberculosis in the United States and in the world.

The use of immunization against tuberculosis infections on a wide scale has never really been attempted in the United States, although other countries have obtained encouraging results with the method. Immunization against this disease needs, and is receiving, more study, and in time may become a useful procedure that will help in control of the spread of tuberculosis.

A common childhood disease, whooping cough, is caused by the bacterium *Bordetella pertussis*. This infection has a built-in system that almost guarantees its successful spread from an infected individual to a susceptible individual. As indicated by the common name, a major symptom of the infection is a violent cough. While it occurs most often during childhood, it is not unknown in adults. The infection may be extremely serious in infants, and therefore, early immunization in the form of DPT shots is strongly recommended. Such immunization is routinely used in the United States, although there is some controversy about its absolute safety. The immunity produced is effective and with proper booster injections may last almost indefinitely. While immunization is the most effective means for control of the disease, again the simple habits of personal hygiene and good sanitation in the presence of the disease will reduce the spread of the bacterium and thus, the potential of transmission to large numbers of individuals. Because whooping cough (or pertussis) is largely a childhood disease, it is particularly a problem when a case occurs in a school environment. The infected child should be removed from school and remain out as long as laboratory tests indicate that the organisms are present in the respiratory tract. The diagnosis is simple, needing only a culture of a cough plate, exposed to the respiratory discharges of the child.

At one time a fairly common disease among children in the United States was diphtheria. The incidence of the disease has been reduced in recent years because most children are immunized early in life as part of a routine procedure in the DPT shot. Diphtheria is caused by *Corynebacterium diphtheriae*. This organism localizes in the throat around the tonsils and occasionally in the nose and produces a pseudomembrane. This membrane, in itself, is bothersome and if formation continues may actually cause suffocation. Much more dangerous is the fact that, living in the throat, the bacteria produce a powerful toxin that is absorbed into the bloodstream and circulates throughout the body producing a systemic poisoning. This toxin causes damage to the heart muscle, to nerve tissue, and to the kidneys. In treatment of the disease, antibiotics are used to stop or prevent the growth of the organism in the throat, and at the same time antitoxin is used to neutralize the soluble toxin. In spite of immunizing procedures, cases still occur. When they do, it is important that the patient be isolated and that this isolation be continued until cultures demonstrate that the person is free of the bacterium. The organism is transmitted from person to person, either by infected cases or by healthy carriers, and in contaminated milk. It is transmitted in discharges of the upper respiratory tract, either directly or indirectly. A skin test, the Schick test, may be used to detect susceptibility to the disease; susceptible individuals should be immunized. Again, isolation or quarantine should be enforced in the presence of any case, and susceptible individuals who have been exposed should be immunized. Prompt and thorough disinfection of all personal articles, particularly of upper respiratory tract discharges such as sputum from coughing or sneezing; of dishes; and any materials used by the patient should be practiced in cases of diphtheria. With such precautions, and with continued education urging immunization, this disease possibly could be eliminated from our environment.

There are a number of diseases produced by viruses that are transmitted by discharges of the upper respiratory tract. Among these the common cold is, without doubt, the disease of highest incidence. The cold appears to be almost an integral part of civilization, with human susceptibility being greatest in the very young, and diminishing with age. It is believed today that several viruses, all of which have not as yet been completely characterized, cause the condition referred to as the common cold. The disease is spread by droplets of discharges from the respiratory tract. This droplet spread may be direct, or indirect by way of articles freshly contaminated. The incubation period appears to be about twelve to forty-eight hours, during which time an acutely irritated condition of the nose, throat, sinuses, and sometimes the trachea and bronchii develops. This irritation lasts from two days to one week. The infection may be accompanied by some rise in temperature, generally not too great, and by a general run-down feeling. In perhaps most of the cases, the period of considerable irritation (sneezing, coughing, and drainage of watery mucous from the upper respiratory tract) may be followed by a secondary bacterial type of infection in the sinuses, throat, or the chest. It is in this secondary bacterial infection that antibiotics may be useful. Antibiotics are ineffective in the early stages, although some of the antihistaminic preparations appear to be useful to relieve the local nasal congestion in some persons. There is no effective immunization. The only means of prevention of the common cold consists of good personal hygiene and sanitation, such as disinfection of personal articles, eating utensils, clothing, and particularly sanitary disposal of tissues used to catch discharges of the upper respiratory tract.

In the past several years a group of viruses has been identified, and studied, which has been shown to be associated with disease of the upper respiratory tract: the Adenoviruses. Symptoms of infection may be similar to those of the common cold; however, some of these viruses are known to produce more acute and dangerous respiratory diseases, including atypical pneumonia. Some types of viruses are also able to infect the membranes of the eye, producing a conjunctivitis. For two virus types a vaccine has been produced which seems to be effective; however, vaccine use is not widespread for viruses in this group.

Influenza, or flu, is a common viral disease of the upper respiratory tract. Generally speaking, it is relatively mild and in some rare cases may be almost indistinguishable from the common cold. In most cases, however, although the virus infection itself may be short-lived, it produces a debility, general lassitude, and a run-down condition that may last for weeks. Influenza in epidemics may also be a major contributing cause of death, particularly in the elderly. In most cases death results from a bacterial pneumonia which frequently follows the influenza infection. The highest mortality rates during epidemics are in persons over 40 years old, and those over 60 are a particularly great risk because of frequent complications. The influenza virus is transmitted either by direct contact or by indirect transfer of upper respiratory tract discharges. The virus may survive for some time on articles freshly soiled with nasal discharges, and indirect transfer may be fairly common. After an incubation period of one to three days, the disease appears suddenly, with irritation of the nose, throat, and upper respiratory tract. Symptoms may include headache, muscle pain, sore throat, and frequently some discomfort or complication in the lungs. Most characteristic is the production of marked weakness caused by increased temperature. This weakness is almost always followed by aches and pains throughout the body. The actual infection rarely lasts more than a few days; however, the patient is weakened and made much more susceptible to bacterial pneumonia following an influenza infection. While antibiotics do not influence the course of an influenza infection, they will help to prevent the secondary bacterial pneumonia and are often used to treat influenza for this reason. Immunization against influenza is said to be from 60% to 75% effective in preventing the disease. It is particularly important that people with chronic disease and those in the most susceptible group, over 60 years of age, should be immunized. Immunization procedures vary depending on vaccine and on the strain of virus that is expected in a community in any given year. Generally a yearly booster, or injection, is necessary to provide adequate protection, because the virus causing an epidemic may change from year to year. In prevention of influenza, good personal hygiene and sanitary practices such as disinfection of soiled articles and disposal of respiratory tract discharges may be effective.

BIBLIOGRAPHY

AMERICAN PUBLIC HEALTH ASSOCIATION. 1980. A. S. Benenson, ed. *Control of communicable diseases in man*. 13th ed. New York: American Public Health Association.

INTERNATIONAL ASSOCIATION OF MILK, FOOD, AND ENVIRONMENTAL SANITARIANS, INC. 1987. Procedures to investigate foodborne illness. 4th ed. Committee on Communicable Diseases Affecting Man (eds.). Ames, IA.

JAWETZ, E., MELNICK, J. L., and ADELBERG, E. A. 1984. *Review of medical microbiology*. 16th ed. Los Altos, CA: Lange Medical Publications.

ROBERTS, H. R., ed. 1981. *Food safety*. New York: John Wiley & Sons.

TROLLER, J. A. 1983. *Sanitation in food processing*. New York: Academic Press.

U.S. DEPARTMENT OF HEALTH, EDUCATION, AND WELFARE. Food and Drug Administration. 1976. *Food service sanitation manual*. Washington, DC. DHEW Publication no. (FDA)78-2081.

WEISER, H. H., MOUNTNEY, G. J., and GOULD, W. A. 1971. *Practical food microbiology and technology*. 2nd ed. Westport, CT: AVI Publishing Co.

Control of Microorganisms

The nature and special characteristics of groups of microorganisms have been considered in earlier chapters. In this section the major emphasis will be on the interrelationships between microorganisms; between microorganisms and the human; and between microorganisms and their environment. The environment of the microorganism includes all chemical, physical, and biological factors that are near enough to affect the functioning microbial cell. The biological factor is of major importance to us because we make up at least a portion of the biological environment for these cells.

In every location in the natural environment that has been studied, some microorganism has been found to exist or to be able to grow. Because of this ubiquitous nature, it is impossible to limit the location of microorganisms in a general manner; rather, the aim is for specific control of certain groups of microorganisms. Control must be managed in such a way that the microorganism is made to benefit humanity. Microbial populations must be limited in such a way that there is no detriment to humans, or be completely removed from the environment of humans. Control does not always mean the complete elimination of microorganisms. In some situations of control we actually work at stimulating microorganism growth. For instance, in using microorganisms to prepare food and to produce chemicals, we seek to stimulate the growth of microorganisms and to increase those activities that we deem desirable. On the other hand, when microorganisms are present that cause some difficulty for us, we attempt to remove them or at least to reduce their numbers. For example, in attempting to prevent food spoilage, we try to eliminate microorganisms; and to prevent certain diseases, we attempt to remove certain portions of the microbial populations. We attempt to eliminate or to reduce greatly the numbers of microorganisms that cause disintegration of textiles or wood products. In those cases where we can control our environment, we attempt to eliminate, or to greatly reduce the numbers of microorganisms that can cause disease in humans or in domestic animals.

The last example of control, in which we attempt to prevent disease, and also in some cases when we attempt to prevent food spoilage, describes the science of sanitation. Such control may be accomplished by reduction of numbers of microorganisms, or in some cases, effectiveness requires that we totally eliminate certain organisms. Because of the different requirements, the means of control that will have the desired effect must be selected for use (see Figure 4.1).

STERILIZATION

Sterilization is a means of control that eliminates all forms of living material. Because of this, sterilization can be applied only to inanimate objects and spaces. It cannot be applied to the human body, because it would also eliminate the living cells of that body. There are several ways to sterilize, although the most common is the application of heat. The degree of heat, or the temperature attained is the important factor in heat sterilization. Heat kills as it coagulates or denatures the proteins of living cells, thus inactivating enzymes and destroying essential structures and functions. The type of heat used, that is, moist or dry, will vary depending upon the objects to be sterilized, and upon the time and temperature that can be tolerated by those objects without damage. Generally, moist heat requires less time than dry heat.

Dry heat is usually applied to objects in a hot air oven. Sterilization with dry heat requires a temperature of from 320° to 356° F (160° to 180° C) for a period of one to two hours. Dry heat is effective for sterilizing glassware, instruments, and objects that moisture might damage. Dry heat is not satisfactory for sterilization of materials containing water or for materials that break down under prolonged periods of heating. Objects can be wrapped in paper, sterilized by dry heat, and when sealed, can be maintained in a sterile condition for a relatively long period of time.

For some sterilization processes, moist heat is more satisfactory. This may be applied by use of steam or boiling water, but most effective is the application of steam under pressure. Steam is applied under pressure in the ordinary pressure cooker or autoclave. By increasing the pressure the temperature is increased, and with the higher temperature the application time is reduced, so that materials that are adversely affected by heat applied over long periods are not damaged.

Boiling water will kill cells; however, the time required to kill spore-producing microorganisms is extensive, and the practicality of boiling is reduced because of the resistance of such spores.

Fig. 4.1.   Mechanical dishwashing system.

Another common means of sterilization of liquids is by use of bacterial filters. A bacterial filter is made with pores so small that the smallest bacterial cells cannot pass through. If a fluid is passed through such a filter, the cells present are trapped on the surface and the fluid removed from the opposite side is sterile. If collected in sterile containers, it will remain sterile.

Ultraviolet light when properly applied is an effective means of sterilizing some materials that would be harmed by heat, and that cannot be filtered. Ultraviolet (UV) light has a wavelength that is able to penetrate cells and to disrupt normal function. The penetrating power of UV light for many substances is not great. It does not pass through soft glass and will not penetrate liquids to great depths. The effectiveness and penetration of UV light depends upon the closeness of the light source. The closer the source, the more effective the killing. This is an especially effective means of sterilization of small volumes of liquids in plastic containers that will not withstand heat even for short periods of time; for example, the use of UV light in industry to sterilize plastic medicine containers is quite common. Ultraviolet light is also widely used in ducts of ventilating systems, either heating or air-conditioning. It is quite effective in reducing the possibility of airborne infections in hospital operating rooms. Ultraviolet light is also used in barber shops and beauty salons for sterilizing combs, scissors, and clippers, and if applied long enough is quite effective.

PASTEURIZATION

In the previous discussion, methods of completely removing all life forms have been discussed for limited, specific, environments. For some applications to foods, and for application to the body, it is necessary that we use methods to reduce the number of organisms rather than completely eliminating them. Perhaps the oldest and most widely used method for liquids is pasteurization. Pasteurization of milk and other liquid foods is a routine process. It consists of raising the temperature of the product to a temperature below boiling for a short period of time, thus killing the more sensitive vegetative cells (microorganisms) that may be present; or to higher temperatures for extremely brief times. Pasteurization does not and was never intended to sterilize. It is intended to remove those cells that might cause infectious disease in man. We use it primarily for milk and food products. It was originally developed for use in the wine industry of France to eliminate the bacterial cells present while permitting the yeast to live and to produce the desired fermentation. Currently it is used in brewing and wine industries as a means of controlling the kind of microorganisms present in these products.

ANTIBIOTICS

In the past thirty to forty years, a different kind of control agent has
been developed for controlling microorganisms, primarily in treatment
of disease after it occurs. These antimicrobial agents may be synthetic
chemicals or naturally occurring chemicals produced by other living
organisms. The latter category includes the antibiotics. When first introduced toward the end of World War II, these were called wonder
drugs, and it was widely believed that most of our problems with infectious diseases were in the past. Unfortunately, this belief was not
well founded, and we continue to have great public health problems
with infectious diseases of all types. While the antimicrobial chemicals,
either synthetic or natural, will help to control and limit the populations
of microorganisms, they are expensive enough and their action is specific enough that it is not wise to use them to control microbial populations in the general environment. Rather, they are more properly
and more profitably used to treat disease. Mankind has had unpleasant
experiences in assuming that because these wonder drugs are available,
it is no longer necessary to properly use the ordinary antiseptics, disinfectants, sanitizers, and germicides. These chemical control agents
will reduce the chance of infection, and those infections that do occur
in spite of sanitary procedures can then be treated with the antimicrobial agents specific for certain disease organisms.

Following the discovery of antibiotics and therapeutic drugs for
many protozoan diseases, a widespread practice has emerged in animal
husbandry of adding these medications to animal feeds, which has resulted in promotion of animal growth. The mechanism of the antimicrobial substance in promoting growth is not fully understood; however,
this practice may result in problems ranging from inducting drug-re-
sistant microorganisms, to causing allergic or toxic reactions in the
human. For these reasons this practice is receiving a much closer look,
both by governmental agencies and scientists, and is now much more
closely regulated. Certain foods have had allowance concentrations of
antibiotic substances specified. On the other hand, cleaning compounds,
disinfectants, and so on are specifically controlled by laws and operating
requirements because many are toxic to humans. The conditions of
storage, use, and the prohibition of residues of any detectable amounts
in foods are specified.

CHEMICAL AGENTS

In many cases we need a method to control microorganisms in, on, or
around the human body. For this purpose, we have many chemical
agents that are useful in helping us to control numbers of microor-

ganisms. These agents have been given various names that originally had some significance as to the type of action desired, for example, antiseptic, germicide, disinfectant, sanitizer, and so on. The technical difference between these terms has been somewhat obscured by their indiscriminate use, regardless of the action desired. An antiseptic, for example, is a chemical agent that will prevent sepsis or infection. This means simply that it will reduce the numbers of microorganisms to a level below that required for the production of infection.

Germicide is a general term applied to a chemical that will reduce the number of microorganisms in a environment without particular regard to the nature of these organisms; that is, whether they are pathogenic or not. A sanitizer is much the same type of chemical agent as a germicide.

A disinfectant is a chemical that can be applied to objects or the human skin that will kill growing forms, but not necessarily the resistant forms, of disease-producing microorganisms. Many disinfectants are too strong and too toxic for application directly to the human skin, while others are mild enough so that application to living tissues does not produce injury. Actually, many soaps are mild disinfectants. Concentrated Lysol® or phenol are too strong for direct application to the human body and cannot be used routinely. Both of these, however, are disinfectants because they will destroy the growing forms of infectious disease organisms. Different brands of disinfectant chemicals may be tested for effectiveness by comparison with phenol to determine the phenol coefficient as follows:

(1) Use either *Salmonella typhi* or *Staphylococcus aureus* specified strains for the test.

(2) Compare effectiveness of chemicals to phenol on the basis of concentration (dilution) and time as given in Table 4.1.

(3) If chemical (X) allows growth in a 1:150 dilution at 5-min. exposure, but not at 10-min. exposure, and if phenol allows growth in 1:90 dilution at 5-min. exposure but not at 10-min. exposure, then the phenol coefficient is

$$\frac{150}{90} = 1.66$$

and chemical (X) can be used with confidence at a dilution 1.66 greater than the required dilution of phenol.

No chemical presently known can be said to be the ideal disinfectant for any and all purposes. However, a disinfectant whose use is desirable must have certain characteristics. These properties would include

(1) toxicity to disease organisms, meaning capacity of the chemical to kill microorganisms that can produce disease in man;
(2) solubility (i.e., it must go into water solution so that it can be applied conveniently to objects in the human environment or possibly to the human);
(3) stability (i.e., it must not break down or become ineffective upon standing);
(4) lack of toxicity to humans or to domestic animals;
(5) activity or structure not greatly affected by changes in temperature;
(6) non-corrosive, and not capable of staining surfaces to which it is applied;
(7) cleaning capability;
(8) odorless, or as nearly odorless as possible and able to help eliminate undesirable odors that may be present in the environment; and
(9) readily available at low prices.

If the disinfectant is toxic, this property can frequently be reduced or removed by dilution of the chemical unless the disinfectant action of the chemical is also reduced accordingly. Because disinfectants must be used in a variety of places and circumstances, it is important that the disinfectant action be available at low or high temperatures (i.e., in summer or winter). It is obvious that any chemical that leaves any discoloration will not be used more than once; nor will a chemical be reused if it is irritating to the skin. If the disinfectant acts as a detergent

Table 4.1. Phenol Coefficient Table

Concentration (Dilution[1])	Time of Exposure (min.)		
	5	10	15
Chemical (X)			
1:50	No growth	No growth	No growth
1:100	No growth	No growth	No growth
1:150	Growth	No growth	No growth
1:200	Growth	Growth	No growth
1:250	Growth	Growth	Growth
1:300	Growth	Growth	Growth
Phenol			
1:70	No growth	No growth	No growth
1:80	No growth	No growth	No growth
1:90	Growth	No growth	No growth
1:100	Growth	Growth	No growth
1:110	Growth	Growth	Growth
1:120	Growth	Growth	Growth

[1]Dilution expressed as 1 part chemical plus 49 parts water.

in addition to its killing capacities, then it accomplishes two purposes and is more likely to be used in a variety of locations. Most disinfectants are sold in a concentrated form so that they can be diluted for use under ordinary circumstances, and if they can approach the ideal characteristics just listed in this kind of use, then they become more likely to be used with frequency in all sorts of circumstances. Unfortunately, few chemicals possess all the ideal characteristics; therefore, we choose the best available.

What are some of the chemicals that make good disinfectants? Joseph Lister, in the 1860s, applied carbolic acid to hospital operating rooms and showed that the result was a reduction in danger of infection in the surgical patients. Carbolic acid, or phenol, used by Lister is still a popular disinfectant, although the original compound has some undesirable characteristics. Chemists have worked magic with phenol by changing the formula only slightly with addition of other chemical groups and have managed to make phenolic compounds that are more desirable as chemical disinfectants than phenol itself. The activity of phenol as a chemical, however, makes it good for use as a standard for reference in disinfection. A test is easily run to compare any new disinfectant to phenol, as described earlier in reference to the phenol coefficient. The results of such a test readily tell the user whether the disinfectant is better or poorer than phenol for killing disease organisms in the environment. If the phenol coefficient of a compound is greater than one, then it is more efficient than phenol for killing disease organisms. Obviously the higher the number applied as the phenol coefficient for a compound, the better the killing power and the more desirable for use as a disinfectant; provided it has the other characteristics mentioned earlier.

Other chemicals that may make useful disinfectants include various kinds of alcohols, iodines, chlorine and chlorine compounds, heavy metals in various chemical combination, dyes, soaps, and quaternary ammonium compounds.

CONTROL OF MICROORGANISMS IN FOODSERVICE ESTABLISHMENTS

In a food preparation area in a service establishment, it is essential that the amount of microbial contamination in the environment be kept at a minimum. In order to attain this goal, regular, routine, thorough regimes of cleaning must be constantly practiced and followed by disinfection. For this purpose, there are chemical compounds that will do the best job of cleaning and disinfecting different surfaces in the preparation area, and will leave the minimum amount of residue. For example, hand surfaces in such an environment must always be cleaned, and the soaps and disinfectants used for the hands will not be the same as those used for glass, ceramics, stainless steel, plastic, or fabrics. Some chemicals will be too strong or abrasive for routine use in some locations, but others are available for regular, frequent cleaning and disinfecting of all surfaces in the areas. Although soaps and detergents are disinfectants, they are weak disinfectants, and considerable time is required before they can accomplish much killing of microorganisms in any environment. In the use of these chemicals, the agitation, scrubbing, or rubbing will be more effective in removing microorganisms than will the mere presence of the soap or detergent. In food preparation areas, it is important that neither cleaners nor disinfectants be left as residues on surfaces or in appliances where foods may be contaminated. For this purpose, the more water soluble the chemical, the better for removing residues in the rinsing operation.

The major importance of soaps and detergents in cleaning food preparation areas is that these compounds help to emulsify and remove food residues from surfaces, but some agitation must be applied to help them. The removal of food residues reduces the potential for growth of microorganisms, and when such growth is controlled, the amount of contamination passed from one food preparation to the next is held to a minimal level.

CONTROL OF CONTAMINATION DURING PROCESSING

Processing of food of any kind should consist of a series of steps toward making the food cleaner, more desirable, and more wholesome. Unfortunately, in processing, foods must be handled by the human, machines, or both. Humans must build, operate, and maintain the machines, and because these humans, at best, are likely to be carrying organisms that may produce infection, the foods may become contaminated by contact with these humans, machines, or both. Such contamination may occur even when food processors think they are being clean, unless special precautions are taken. These precautions include

(1)   preventing other animals (insects, rodents, or birds) from coming in contact with foods or any food preparation surface at any time;

(2)   providing protective clothing for workers to protect themselves, and to protect the foods from contamination from workers' street clothing;

(3) keeping machinery, tables, sinks, and rooms clean and uncluttered to prevent the accumulation of filth in the form of
spoiled or stale food scraps and wastes;

(4) removing wastes, trash, and garbage from processing rooms
and disposing of them properly to avoid attracting insects and
rodents;

(5) keeping storage facilities clean, uncluttered, and properly refrigerated to prevent cumulative contamination and spoilage;

(6) keeping machinery and instruments in good working order to
prevent breakdowns and delays;

(7) providing adequate and clean dressing, washing, and restroom
facilities for workers to allow the workers to control contamination during the work period;

(8) providing health tests and examinations of employees on a
timely and regular basis;

(9) cooperating with government inspectors so they can observe
the entire operation;

(10) making every effort to correct each fault noted when any inspection report is received.

Processed food should be fresh, clean, and free from putrid or poisonous materials and from anything that can be termed filthy. Fresh
foods will contain fewer microorganisms if treated in the manner described earlier. In most large establishments it is both necessary and
less expensive if raw materials are purchased in large quantities. This
practice may, however, require that raw materials be stored for relatively long periods before use. Storage of canned foods may be accomplished easily as concerns sanitation; however, the space should be kept
clean and uncluttered in order to prevent attraction of insects and rodents. Adequate storage space may also prevent overlooking certain
stocks for long periods, which would permit corrosion of containers and
possible loss from spoilage.

Storage of fresh or frozen foods presents greater problems in sanitation. If refrigeration is required, it is essential that adequate space
be available for air circulation to permit proper cooling, and that sufficient space is available for moving supplies to permit thorough cleaning of refrigeration space. Generally, it is required that foods needing
refrigeration be kept at 45° F (7° C) or below. This, in most cases, means
that food may not be removed from refrigerated spaces, even for short
periods to permit cleaning. Removal from such space for 30 minutes
to one hour is sufficient to permit one or more generations of bacteria
to be produced. When the food is placed back inside the refrigerated

space, additional time is necessary to reduce the food temperature back to the required level. Repeated exposure will permit production of a number of organisms sufficient to constitute a hazard in terms of contamination that may be carried to the consumer. At the least, there may be alterations in the flavor or quality of the food if it remains at the higher temperature for any appreciable time.

Cleaning of refrigerated spaces should be as thorough as cleaning of any other space, equipment, or facility for food preparation. Not only should any obvious dirt, trash, or food scraps be removed, but adequate disinfection, without danger of chemical adulteration, should be accomplished. Many microorganisms are capable of growth at fairly low temperatures. Low temperature microorganisms, or psychrophiles, may cause complications for food industries, varying from food spoilage to toxicity for the consumer. In most cases, disinfectants will kill psychrophiles with effectiveness equal to that for mesophilic bacteria. The removal of food scraps, trash, and dirt reduces the possibility of growth of microbial populations present on foods when they arrive at the storage facility.

Different cleaners and chemical disinfectants should be selected for use on different kinds of food-contact surfaces and equipment. When the temperature can be raised to 170° F (77° C) or above, water alone is preferred over chemical disinfectants for any kind of surface. Below that temperature, alkaline, nonionic detergents are preferred for cleaning glass, ceramic ware, stainless steel, plastics, rubber, wood, paint, and fabric surfaces. Iodine, or quaternary ammonium compounds may be used to disinfect all except fabrics. Hypochlorites or organic chlorine compounds may be used to disinfect all the above, including fabrics. Floors, walls, windows, and any non-food contact surface in the work area can be adequately cleaned with a general purpose cleaner, and disinfected with a good, non-staining disinfectant. Such general cleaning and disinfecting is important to the overall sanitary condition of the food handling environment. In heavy contamination areas, particularly those involving grease, special scouring powders or solvents may be necessary to remove residues that occur in sinks or on some cooking utensils. Any cleaning compound or tool that is greatly corrosive or abrasive should not be used. Wherever possible, disposable paper, cloth, and sponges should be used in preference to non-disposable materials, because scrubbers made of these materials are almost impossible to maintain in a disinfected or sanitary state.

Generally, two separate freezer storage spaces are desirable rather than one large one in order to completely empty each for periodic cleaning and upkeep, and also for emergency situations when one facility fails. Storage space for frozen foods must be adequate for rotation of stock. The stock rotation permits checking inventory to avoid storage for periods so long that foods lose flavor and quality. Because frozen foods cannot be removed from freezing to room temperatures or refrigerator temperatures for even short periods without some degree of thaw, the rotation of stock in freezer space is essential. Thawing frozen foods to any degree permits possible growth on the surfaces of microbial populations present. This increase in population may cause spoilage, production of putrid or poisonous substances, and may increase the danger of food poisoning or infection to the consumer. At the very least, any degree of thaw, followed by refreezing will result in loss of juices and flavor from the food.

Foods in dry form, not needing refrigeration, present different problems from the sanitary viewpoint. Storage areas for foods of this type (examples: cereals, flours, etc.) will be particularly attractive to insects and rodents. In some instances, special insect- or rodent-proof storage facilities are desirable or required. For food amounts that can be handled easily, refrigeration is a good means of keeping insects away from dry foods such as those just mentioned. Adequate precautions are needed to prevent invasion of such facilities by these animals. The presence of insects in storage facilities will mean reproduction of the insects and adulteration of the foods by the presence of eggs or larval forms. In addition, the adult insects may bring microbial contamination to the foods. The presence of rodents in a storage area may result in actual destruction of food, and certainly will result in the contamination of food by microorganisms from the urine and feces of the animal. Storage spaces for dry goods should permit rotation of stocks to maintain freshness, and should be of sufficient size that stocks can be moved for complete cleaning. No materials other than foods should be stored in the same space. All boxes, papers, and trash should be kept out of such spaces to remove potential hiding places for insects and rodents. The use of insecticides must be judicious to avoid the adulteration of foods by these chemicals. In general, it is preferable to use insect- and rodent-proofing together with frequent cleaning instead of indiscriminate use of chemical poisons.

In summary, food storage spaces should be kept at least as clean, uncluttered, and sanitary as food preparation spaces in order to prevent the addition and actual cultivation of contamination in stored foods. Foods that were good and wholesome on delivery may be made unfit for human consumption by storage in inadequate, dirty spaces even for a short time.

If good quality, uncontaminated foods are taken out of storage for preparation, then the job of the cook is made much easier and the end product is much more likely to be wholesome and palatable. Yet at this stage, the most rigid sanitation practices are more desirable, or at the least cannot be relaxed. Foods spoilage or pollution cannot now be reversed; but it *is* possible to prevent the addition of contamination from this time until the food is consumed.

To accomplish this end, it is essential that the food enter a clean, uncluttered, disinfected preparation space, the kitchen. Many small spaces are perfectly satisfactory for food preparation, if the space is well planned and is used only for food preparation. To accomplish this, all surfaces should be easily cleanable, with no seams where food particles may collect. These surfaces should not react with, nor give off, harmful chemical substances. It is necessary that such a kitchen be adequately lighted and ventilated, that it be cleaned frequently during food preparation, and that it be thoroughly cleaned and disinfected daily. For this to be possible, all surfaces must be free of stacked materials not absolutely essential to food preparation. This is true of countertops where food may actually come in contact with the surface, and it is true of floors and subcounter surfaces. Permitting the accumulation of trash or food scraps in any place in a kitchen is an open invitation to invasion by insects and rodents with resulting increased danger of pollution or contamination of any food prepared in that space. Because by nature, the kitchen temperature is likely to be somewhat higher than that of serving areas, this increased temperature permits and even encourages the growth of food spoilage and disease-producing organisms that may have come in on some food, or that may have been brought in by a cook or other kitchen worker.

Food preparation surfaces should be located well away from garbage collection or scrap depositing areas, to prevent the accidental entrance of this refuse into any food in preparation. Scraps and garbage should be collected in containers that have been cleaned and disinfected before each use, and should be emptied or removed from the kitchen at frequent intervals during the day. The containers should be of sufficient size that there is never an overflow, but small enough that frequent removal is essential. Garbage and scraps of foods at warm room temperature will rapidly spoil and serve as a breeding ground for most bacteria and fungi. This spoilage is (a) attractive to insects and rodents, (b) a potential source of microbial contamination of foods and workers, and (c) at the very least, most unattractive in appearance and odor to anyone present in the kitchen.

In order for the kitchen to be kept in the condition just described, an abundant supply of hot and cold water is an absolute necessity. These supplies must be available to the cook in a location away from that used for cleaning utensils and dishes. This availability permits cleaning many foods before preparation as well as cleaning food contact surfaces, and even the hands of the cook. These frequent cleaning operations, in all cases, can only result in more wholesome, less contaminated food products.

Facilities for washing and cleaning cooking utensils may be the same as those for cleaning serving dishes and eating utensils, provided that this dual use does not overload the facility. It would be preferable that these washing facilities be separate. This separation would more likely result in preparation utensils being washed and cleaned both more thoroughly and more often. These utensils cannot usually be subjected to chemical disinfectants as used elsewhere. They must, however, be disinfected, and this can be done quite effectively by (a) use of scrubbing or water agitation in automatic equipment, (b) use of adequate soaps or detergents, and (c) use of water of sufficiently high temperature. In the latter respect, properly functioning automatic washing equipment is preferable because of the higher water temperatures, which may be used for washing and for rinsing, and a drying cycle, which can essentially sterilize the utensils (see Figure 4.2). Hand drying should not be used even with sterilized drying towels. And it is useless to wash and disinfect a cooking vessel only to wipe it dry with a dirty, contaminated towel.

Foods placed in thoroughly clean utensils must still be properly cooked in order to be safe for consumption. Common cooking temperatures for most foods are perfectly adequate for killing most vegetative cells of bacteria, rickettsia, protozoa, and fungi, and for inactivating viruses that may be present in food. These temperatures are usually not adequate to kill bacterial or fungal spores, and cannot be depended upon for preservation of foods for long periods of time without refrigeration. For effectiveness, any temperature used in cooking a food properly must be matched with the proper time as well, to kill or inactivate microorganisms (see Figure 4.3). Cooking temperatures are those higher than that required to prevent growth of microorganisms. This minimum temperature is considered to be 140° F (60° C).

Once cooked, many foods are often allowed to stand at least a few minutes at room temperature before reaching the consumer. This period should be short. Foods removed from cooking temperature cool to levels still warm enough to encourage growth of many bacteria. Any accidental contamination by a worker at this stage, therefore, will provide an inoculum for a culture of potentially dangerous microorganisms. Most regulations require that foods remain at temperatures of 45° F (7° C) or below, or 140° F (60° C) or above except during *necessary* periods of preparation and service. This requirement prevents accidental incubation of bacterial cultures in foods.

Many foods are not cooked at all, or are cooked and then refrigerated

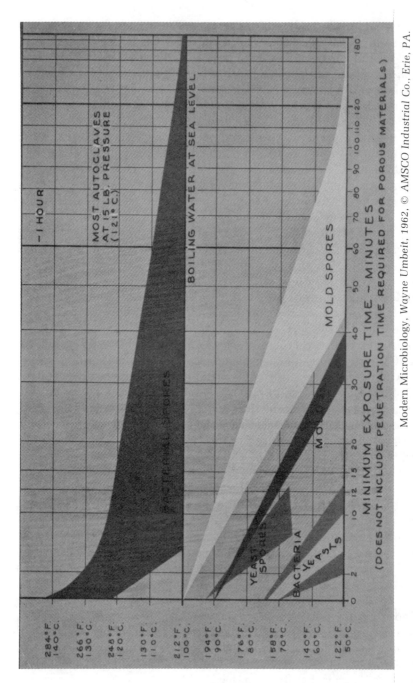

Fig. 4.2. Temperatures required to kill various classes of microorganisms.

Modern Microbiology, Wayne Umbeit, 1962. © AMSCO Industrial Co., Erie, PA.

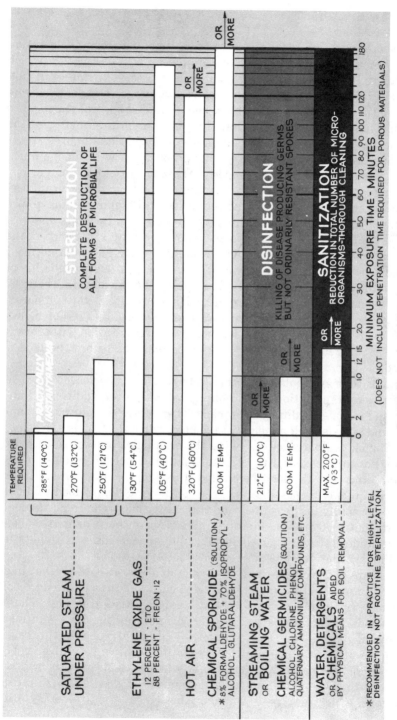

Fig. 4.3. Methods for controlling microbial life.

© 1967 Research and Development Section, American Sterilizer Co., Erie, PA.

103

to be served cold, or possibly cooked and cooled only to room temperature before serving. The dangers of contamination of these foods is much greater than in the cases in which foods are served while still hot. Uncooked foods will have some bacterial or fungal populations from their natural states, and will have been handled several times since original harvest. Each step of handling has potentially increased the contamination of the foods. Washing, peeling, and cleaning fresh foods such as vegetables and fruits in salad preparation will remove much of this contamination, if done properly. These processes done in a sloppy fashion may serve to transfer the contamination from the external peel to the interior foods if, for example, dirty fruits and vegetables are handled by peeling before washing. Once peeled, many fruits and vegetables cannot be adequately washed and if washing is attempted, the product may lose much of its food value and flavor. Raw salad foods should first be thoroughly washed. After the cook has washed his or her hands, and the countertop where he or she is working, then the cleaned foods should be peeled, trimmed, and prepared. Once prepared, the raw foods should be refrigerated at once to 45° F (7° C) or below. If any degree of contamination has occurred, refrigeration will prevent any great increase in numbers of microorganisms before the food is served.

Foods that are cooked and then refrigerated to be served cool are subject to contamination after cooking because of the additional handling and transfer to different utensils. These periods required for handling and transfer will generally take place when the food is at a somewhat higher temperature, allowing the organisms present to be most active. For this reason, it is essential that these periods be as short as possible and that the food be chilled (throughout) to 45° F (7° C) or below as quickly as possible. A large container of warm food will not be cooled rapidly in the center, and this center may remain at an ideal growth temperature for microorganisms for an hour or more, even though under refrigeration. It is advisable in such cases to divide the food into smaller portions for refrigeration.

Perhaps the greatest dangers of microbial growth are present in a third situation: that in which foods are cooked and then allowed to cool to room temperature for service, for example, salad dressings. These warm foods, if contaminated, provide optimum incubation conditions for any microorganisms that may have gained entrance. In this case, the small amount of contamination gaining entrance will be compounded many times within one to two hours. Such growth may be sufficient to (a) produce putrid or decomposed ingredients; (b) cause infections, which may require rather short incubation times because of the numbers involved; or (c) produce exotoxins that result in food poisoning in any consumer. If it is necessary to serve food in this way, then it is equally necessary that the food be prepared in small quantities, be refrigerated until shortly before use, and then be warmed only in needed quantities.

In some situations, foods that must be prepared in large quantities and kept warm may be most dangerous of all, especially if the temperature is not kept at 140° F (60° C) or above. Below this point, microbial cells are not killed, and at least some may grow and increase in number. Meats, poultry, and gravies fall into this food category.

The essentials for preventing foodborne diseases, from preparation to serving, may be illustrated in step-wise form:

RAW FOODS   Must be cleaned in potable water—handled carefully

$\downarrow$

Preserved* or Refrigerated—Processed

$\downarrow$

Prepared—Heated to 140° F or Cooled to 40° F or below

$\downarrow$

Leftovers, Refrigerate to 40° F or Freeze—Serve

$\downarrow$

Prepare and serve at once.

---

*May be preserved by heating, addition of spices, addition of salts, addition of sugars, addition of chemical preservatives, drying, or freezing.

In the United States, the word *hygiene* refers primarily to personal cleanliness and personal habits, while in other countries the word may have a much broader meaning. Personal hygiene relates to the attributes of the individual. Careless persons handling food products obviously distribute potential disease-producing bacteria indiscriminately to unsuspecting consumers of the foods. The best and cleanest facilities and equipment are nullified immediately when some people enter the kitchen. This occurs through no fault of planning or mechanical function. It may occur because of the dress, personal habits, or the health of the person. City ordinances, therefore, specify certain requirements for personnel who are employed in commercial food preparation and service.

Recommended ordinance requirements included in the *U.S. Public Health Service Manual* include suggestions for all employees to be required to "wear clean outer garments, maintain personal cleanliness, and conform to hygienic practices while on duty." It is also recommended that persons obviously infected, suspected of being infected, or of being capable of transmitting disease to others should not be allowed to work in food preparation or service. This recommendation is taken by many persons to mean that persons should be required to undergo medical examinations and be certified free of contagious disease before being allowed to work as food handlers. While it would be very good if we could at any instant know that all food handlers were not capable of transmitting a potential disease-producing organism to another person, it is obviously not possible to constantly examine all persons associated with foods and to make such a certification. Because any examination could provide evidence for such certification for only a few hours at most, the practice of requiring health cards for such certification has been dropped by most authorities. If these recommendations could be carried out, then there would be little or no danger of personnel contaminating foods in processing, preparation, or service. The necessity for wearing clean outer garments has been discussed in detail earlier as related to food production and processing. In food preparation and service, the importance of clean outer clothing is probably greater than in production and processing because of the nearness of these activities to the consumer and the resulting fact that fewer treatments remain for reducing the effects of contamination before the food is consumed.

The maintenance of personal cleanliness is a requirement that may be much more difficult to evaluate and control, yet its importance in a foodservice establishment employee cannot be overemphasized. A worker who does not keep himself or herself clean is unlikely to worry about keeping food clean for others to consume. The conformation to hygienic practices while on duty may give the employer the best clue to the personal cleanliness of the employee. Hygienic practices become habit with those who do conform. Such habits will hold, not only on duty, but in the off-hours as well. If an employee cleans his or her hands frequently, and always after handling any object that may be contaminated, an employer could be relatively certain that that person would also maintain personal cleanliness in other situations as well. Most municipal ordinances require that all personnel wash hands after using the toilet and before returning to work, and recommend hand washing at additional frequent intervals. This requirement, if followed, would prevent much of the spread of infectious disease, particularly those infections of the digestive tract such as salmonellosis or shigellosis.

Persons employed in foodservice establishments, like anyone else, are subject to infection. When communicable infections occur in these persons, they become potentially dangerous to any person with whom they come in direct contact, and to any person who comes in contact with any object previously handled by the one infected. Such an employee of a foodservice business becomes dangerous to all co-workers and to anyone patronizing that business. In earlier chapters, the difficulties involved in such situations were mentioned; yet, to prevent the transmission of infectious disease, it is essential that anyone with an active infection not be allowed to prepare or serve food. Each dish prepared and each dish served is a potential vehicle of transmission of disease. If the infection is located in the respiratory or breathing system, then the employee is potentially dangerous if he or she is present in the food preparation or service areas because of potential transmission by coughing or sneezing. Such situations should be handled by sending the employee to a physician and requiring him or her to stay off duty until the contagious period is past. If this requirement were adequately enforced, the transmission of disease through foodservice businesses would be dramatically reduced.

A more difficult problem is posed when an employee is a carrier of a pathogenic organism and shows no symptoms of infection. As stated previously, this condition can be recognized only by adequate and complete health examinations that include cultures for potentially dangerous bacteria. Requirement for such health examinations before employment and at periodic intervals thereafter will help to locate and permit treatment of carriers, but will not be of help in controlling those individuals who succumb to infectious diseases.

Food industry employees should be concerned enough for their occupations that they make every effort to maintain good health, and to recover that health quickly when infections do occur. Some infections may appear so minor, however, that it may not occur to an employee that he or she represents a danger to another person. For example, a small infected pimple on the face or arm appears relatively harmless. Yet, if it is present on a cook and is caused by staphylococci, the danger may be very real. The cook has only to touch his or her hand to the infection and then touch the food being prepared. The inoculum to begin an epidemic of staphylococcal food poisoning will have been planted.

The employer must be concerned, observant, and sympathetic to employees in maintaining a watch to prevent the spread of diseases from workers to patrons; and must enforce local ordinances fairly and firmly to maintain the business as well as to maintain the health of his or her employees and patrons.

Well-designed facilities, good utensils, careful preparation, and conscientious employees will turn out good, wholesome, attractive food for patrons of a foodservice business. This result will continue if the entire facility is thoroughly cleaned at regular intervals. Thorough cleaning does not mean simply wiping off the countertops and equipment and washing the pots and pans. Thorough cleaning means washing off the countertops completely and disinfecting them. It means complete cleaning of any mechanical device, including the can opener. It means storing utensils in clean, dust-free, and rodent- and insect-free cabinets. It means sweeping, washing, and disinfecting the floor. Thorough cleaning leaves no food scraps or crumbs on any surface to attract insects and rodents or to serve as a culture medium for microorganisms. Only in this situation can food preparation begin without a built-in population of microorganisms waiting to join in the preparations. In all foodservice businesses, some period each day must be provided for thorough cleaning if a sanitary facility is to be maintained.

Up to this point, we have dealt with factors involved in food preparation. All these factors are equally involved in foodservice. Cleanable surfaces, utensils, and buildings must be maintained to avoid contamination of good food that should be served. Although the time involved in service may not be as long as in preparation, the cleanliness of facilities and utensils is essential because at this stage no further preventive treatment (cooking) to eliminate contamination will be used. Personnel must touch utensils and dishes in order to serve, and this almost direct contact with the patron means that disease-producing bacteria, fungi, and viruses may go almost directly to a new host. Therefore, it is again imperative that the server be free of transmissible disease, and the detection of carriers becomes of even greater concern.

In addition to those factors discussed in relation to food preparation, and which also apply to service, there are special precautions to be taken in the service of food to patrons. One of these concerns is the tableware that will come into direct contact with both the food and the patron. Glassware, dishes, and eating utensils require even more thorough cleaning than do cooking utensils, because these will not be subjected to heat when in contact with the food. These dishes and utensils must be thoroughly washed, removing all food particles. To accomplish this, it is essential that dishes have no cracks, crevices, or chips. Such damaged dishes should immediately be discarded, because bacteria and other microorganisms hiding in these crevices are protected from cleaning chemicals, and possibly even from heat. Food particles that may be present in these cracks and crevices can protect or serve as nutrients for the living cells. The complete removal of all food particles from all dishes and utensils is necessary not only for the reasons just mentioned, but also for aesthetic reasons. We all know how disturbing it is to sit down to a meal in a restaurant and find food particles on the fork, or lipstick on the glass or cup. Even though the utensil may have been heated sufficiently to kill all microorganisms, our meal is spoiled.

Dishes should be treated to remove or kill all contaminating microorganisms, or at least those that may cause disease. To accomplish this, most ordinances require that the utensils that contact both food and the patron be subject to a temperature (in water) of 170° F (77° C) for at least one half minute. This heat should be applied after the food particles have been cleaned away. Such exposure will kill vegetative microbial cells, and will thus aid in preventing transmission of disease. If this method is not used, then certain chemical treatments such as use of chlorine or iodine solutions may be used to kill the vegetative cells. If either chlorine or iodine is used, then the concentration must be carefully checked and the exposure time matched with the concentration in order to accomplish killing. In some places, other chemicals may be used, but local ordinances should be checked before such use. Care must be taken to use only chemicals that will not leave a residue on the utensil. A chemical residue amounts to adulteration and may be toxic to the patron.

A most disturbing practice of many food servers is the manner of handling drinking cups or glasses. Such dishes should never be handled by the server anywhere near the drinking edge. The hands are always contaminated to some extent with microorganisms, and handling a drinking vessel by the edge is a certain way to transfer organisms from the servers' hands to the patrons' mouths. Servers' hands should be kept off any portion of a dish or piece of silverware that will contact either food or the consumers' mouths (see Figures 4.4 and 4.5). This can be accomplished so easily, yet it is neglected so carelessly in all too many public eating places.

Another frequent practice that can carry contamination to food occurs when plates have been filled in the kitchen, usually while sitting on a counter or table top. They are then stacked on the servers' arm with one plate resting on the food in the plate below it. Any microorganisms or chemicals picked up on the bottom of the upper plate are promptly transferred to the patron who consumes food from the lower plate. This practice saves time and steps for the server, but transmits microorganisms to the consumer. It is quite likely that it will also eventually lose customers for the business.

A common practice that potentially contaminates food is the practice of filling bread plates or baskets by hand. Most ordinances forbid contact of the servers' hands with food, yet this regulation is ignored repeatedly. Serving tongs or spatulas are so inexpensive and so easily used that there is no need for this contamination to occur.

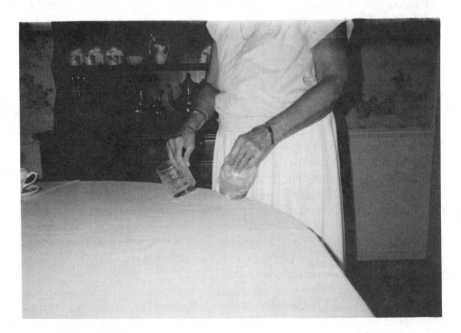

Fig. 4.4. Improper handling of utensils carries contamination from the hands or counters to mouth or food contact surfaces.

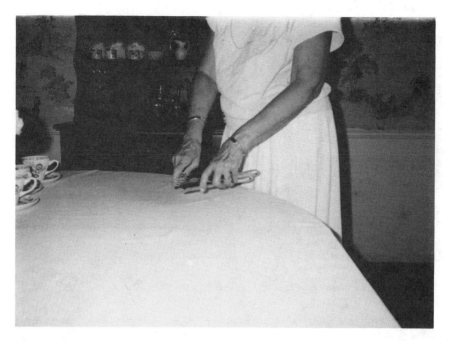

Fig. 4.4. (continued)

Practically all ordinances require that once food is served to one customer, even though it has not been touched, it must be thrown out and never be served to another. The *only* exception to this regulation is for foods individually wrapped, such as crackers or butter, on which the wrapper has not been broken. In any other situation, although the food has not come into contact with any person or eating utensil, it has been exposed to the server, to possible coughs or sneezes and, therefore, respiratory discharges of the customer. To attempt to serve such food again is criminal. This ordinance is frequently broken with re-service of breads from baskets, and with re-service of butter squares—which may or may not have a paper covering, but which are not wrapped. It is also frequently broken by the practice of refilling small crocks or bowls with spreadable cheese or with butter and subsequently re-serving some portion of these foods.

A similar, although permitted situation, is buffet style service of food in public places; it exists also to some extent in some cafeterias. In many such places the foods are placed in cooling trays, warming trays, or at room temperature in large dishes and the customers serve themselves. Hand after hand after hand handles the serving utensils.

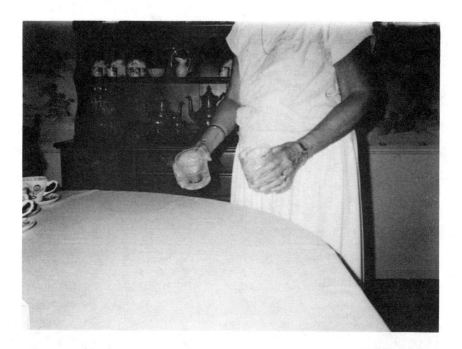

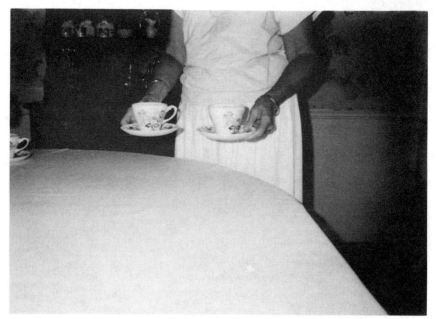

Fig. 4.5. Proper handling of utensils avoids contamination from the hands or counters to mouth or food contact surfaces.

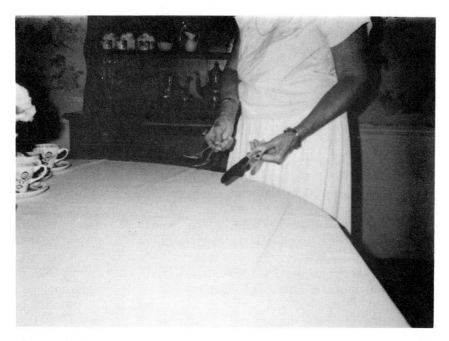

Fig. 4.5.   (continued)

There is no requirement concerning the cleanliness of the clothing or the hands of the person going through the line. How much food goes onto a plate then back into the serving dish when the customer changes his or her mind? Foods sitting at room temperature, or in improperly operating warming or cooling trays provide ample time and adequate growth conditions for most microorganisms, and will at the least hold any contamination that may enter.

Cafeterias eliminate much of this problem by having employees serve the dishes. At least there is chance for less contamination in this way if the employee practices good hygiene. Too often, however, we see a server run a hand across the hair, and continue to serve plates. A woman may fluff her hair, even though it is in a hair net, and continue to serve plates. A hand is rubbed across the face, and the hand returns to serve a plate. A salad dish is filled too full and a hand or a soiled cloth is used to wipe the overflow off the edge of the dish. Countless bacteria or fungi are transferred in each of these actions.

Cafeterias should be required to take precautions against the possibility of contamination by coughs or sneezes, which could be spread over the entire buffet or serving line. This is done most effectively by glass or Plexiglas® shields. Such shields permit the food to be seen by the customer, but prevent some respiratory discharge contamination, and also prevent the customers from handling foods or service utensils.

In any foodservice facility, the cleanliness of all equipment and the rooms is important to prevent food contamination, and is extremely important from the aesthetic or psychological viewpoints. Table and countertops should be *cleaned* (not just wiped with a soiled cloth) after each use. If tablecloths are used, they should be changed frequently, and when washed should be sanitized or disinfected. The room in general should be cleaned regularly. Such cleaning should include windows, light fixtures, air vents, and decorations, as well as floors to remove dust and potential microbial contamination. Cleaning floors at least daily is essential and is sometimes complicated by ordinances that prohibit sweeping while food is being served. Yet in any case, some provision must be made to permit thorough cleaning of all floors and permanent furnishings at frequent intervals. Not to do so permits accumulation of food particles that attract insects and rodents and to provide nutrients for microorganisms.

In food preparation and service, as in production and processing, good sanitation can be maintained by beginning with good quality raw materials, and by routinely using recommended or required procedures as found in most ordinances regulating food industry operations. All foodservice establishments are subject to inspection, sampling, and regulation. Cooperation of owners with local, state, and federal authorities having jurisdiction will help to maintain a foodservice industry of the highest quality, and will help to avoid complaints and possible legal actions by patrons that may result from transmission of disease either directly or indirectly through food. It is advisable for any operator of a foodservice establishment to obtain a copy of all local ordinances applying to the business, and to make these required reading for all employees.

CONTROL OF MICROORGANISMS IN THE PRODUCTION OF FOODS

Plant Foods

Cultivation   (A) Correct procedures and timing must be followed in the use of insecticides and fungicides to prevent crop damage or loss, and to prevent the spread of microbial disease.   (B) Destruction and removal of diseased or damaged plants may prevent spoilage spread to unblemished crops.

Harvest, Transport   (A) Timing of harvest must be prime for the packing and shipping that will be required.   (B) Obviously dirty, or soiled products must be properly cleaned.   (C) All crops must be carefully and properly handled.

Storage   (A) If storage is required, contamination by rodents, birds, or insects must be prevented.   (B) Allow adequate space, air circulation, and provide correct temperature.

Animal Foods

Production   (A) Proper diagnosis, treatment, and separation of diseased animals is necessary. Particular attention must be paid to diseases that are transmissible to the human, or to diseases that may cause decomposition of food products.   (B) Only obviously healthy animals should be used in the production of food products.

Harvest, Transport   (A) Milk must be harvested without added contamination.   (B) Proper refrigeration and rapid transportation must be provided to processing plant for milk or meat products.   (C) Animals must be shipped at prime time, in good transport.   (D) Fish and seafoods must be refrigerated promptly and properly, and shipped rapidly with proper refrigeration.

Storage   If storage is required, adequate refrigeration temperature and space must be provided for the entire time of storage.

CONTROL OF MICROORGANISMS IN PROCESSING OF FOODS

Rodent, Bird, and Insect Proofing

(A) Prevents physical damage to foods.
(B) Prevents added microbial contamination.
(C) Prevents addition of filth to product.
(D) Reduces requirements for use of insecticides, rodenticides, and other chemicals that may adulterate products.

Protective Outer Clothing for Workers

(A) Prevents contamination of products and work rooms.
(B) Prevents contamination of workers from foods carrying microbial contaminants.

Clean Uncluttered Processing Rooms

(A) Do not attract insects or rodents.

(B) Prevent accumulation of wastes that furnish cultures of spoilage- or disease-causing organisms.

(C) Prevent corrosion and damage to machinery.

(D) Prevent contamination of successive process lots from one infected, or infection carrying lot of raw food.

(E) Simplifies and increases efficiency of disinfection.

Rapid and Proper Removal of Wastes, Garbage, and Trash

(A) Prevents the attraction of insects and rodents.

(B) Prevents spoilage of wastes and garbage that may contaminate process lots.

(C) Prevents recontamination in successive process lots.

Adequate Storage Space and Refrigeration

(A) Prevents loss of raw products by overlooking or losing them in storage.

(B) Prevents growth of and contamination by spoilage organisms in storage space.

(C) Permits adequate air circulation for rapid and continuous cooling of storage products.

(D) Permits easier and more efficient cleaning and disinfection.

Adequate Dressing and Rest Room Facilities

(A) Increases attractiveness of work conditions.

(B) Does not attract insects and rodents.

(C) Prevents spreading contamination between personnel.

(D) Reduces chances of carriers among personnel spreading the carried organism to food processing.

Elimination of Infected Workers

(A) Reduces lost-time total for all personnel.

(B) Removes the threat of spread of organisms to process and plant.

(C) Increases well-being and productivity of personnel.

Cooperation with Inspectors

(A) Helps locate potential trouble before it strikes.

(B) Helps locate points of contamination.

(C) Increases the overall quality of products.

(D) Reduces the threat of financial loss from condemnation of products.

Action Following Inspection

(A) Remove potential trouble from any areas indicated.

(B) Prevent financial loss from condemnation of mishandled lots to this point.

(C) Increase overall quality of products, thus increasing the potential profit of the operation.

CONTROL OF MICROORGANISMS IN PREPARATION OF FOOD FOR SERVICE

Food Supplies

(A) Food supplies should be obtained from a source approved by, or considered satisfactory by local or state health authorities. There must be assurance that the food has been processed and shipped in a manner to control or prevent contamination.

(B) Milk or milk products used in preparation, or served fresh must be pasteurized. If cooking is required, dry milk may be used.

(C) Supplies from regularly inspected sources for any type of food are most likely to be uncontaminated and wholesome.

Storage

(A) Sealed containers: (1) Must be stored above the floor to permit cleaning. (2) Must not be subjected to water from leakage from above, or rising water from below.

(B) Packaged or dry foods: (1) Must be stored above the floor. (2) Must be protected from moisture. (3) Must be protected from insects and rodents.

(C) Perishable foods: (1) Must be kept at 45° F (7° C) or below, or 140° F (60° C) or above if not frozen. (2) If frozen, foods must be kept at a temperature to remain so.

Preparation

(A) Foods not requiring cooking must be prepared with a minimum of handling, and a minimum of contact with preparation surfaces.

(B) Raw fruits or vegetables must be washed before being cooked or prepared for service.

(C) Perishable foods should be kept as cool as possible during preparation. If served cold, refrigerate (45° F, 7° C or below) immediately on preparation until service. If served warm, heat to 165° F (74° C) or above and maintain at 140° F (60° C) until served.

Equipment, Utensils

(A) Seamless utensils resistant to denting, chipping, or pitting that can be scrubbed and scoured with cleaning and sanitizing agents are desirable.

(B) Food contact surfaces should be smooth, seamless, and capable of withstanding repeated cleaning.

(C) All equipment, utensils, and surfaces should be cleaned after each usage. Cleansers must be removed after each cleaning so that no residue remains to constitute contamination.

(D) Food preparation surface areas should be free of all articles not required in actual preparation.

Garbage Disposal

(A) Garbage should be kept in durable containers that do not leak. Containers should have tightly fitting covers in place at all times.

(B) Garbage containers should be emptied or removed outside the preparation area at frequent intervals.

(C) Immediately after being emptied, each container should be thoroughly cleaned and disinfected. Containers should be drained and dried before being brought back into the preparation area.

Water and Cleaning Facilities

(A) Water should come from an approved source, and should be provided both hot and cold in unlimited quantities.

(B) Separate cleaning facilities for utensils, equipment, and for hand washing should be maintained.

(C) Toilet facilities should be located convenient to all areas and should be supplied with complete hand washing facilities including adequate soaps and clean towels.

(D) Utensil and equipment washing facilities must provide water of sufficient temperature to provide adequate cleaning. This should be at least 140° F (60° C) for manual cleaning and at least 160° F (71° C) for automatic dishwashers. Rinse water should be at least at a temperature of 180° F (82° C).

Personnel

(A) Should be required to maintain a neat and clean personal appearance and outer clothing should be clean.

(B) Most ordinances require that all food handlers be free of infectious disease. This requirement is not met by the mere requirement of health cards, but must be met by personal responsibility of the food handler and the supervisor.

(C) Periodic, complete, thorough medical examinations should be required by the supervisor in order to attempt to avoid the development of any carrier condition by any food handler, and as a fringe benefit for employees.

(D) Hair nets, head or hair coverings or restraints are required by many ordinances to be worn by persons preparing or serving food. Such restraints will not prevent microorganisms from being removed from the hair, but will prevent loose hair from falling into food.

(E) Prompt reporting of any employee with an obvious infection of any kind, and treatment of this infection by a physician will avoid potential food contamination.

(F) The use of tobacco while preparing or serving food is generally prohibited.

General Cleaning

(A) Thorough, frequent cleaning of the entire foodservice establishment to prevent accumulation of, or to remove dust, dirt, and grime from floors, walls, windows, ceilings, and all permanent installation avoids attracting insects and rodents and prevents buildup of contamination.

Display of Foods for Service

(A) Foods such as pies or pastries, that may be displayed before service are subject to the same regulations as foods in storage. (1) Protect from dust. (2) Protect from insects. (3) Protect from contact with hands or surfaces. (4) Maintain temperature at or below 45° F (7° C) or at or above 140° F (60° C), depending upon nature of food and whether or not it is perishable.

CONTROL OF MICROORGANISMS IN FOODSERVICE

Precautions in Preparation

(A) Care must be continued in service to prevent addition of microorganisms or filth to food after it has been cooked or otherwise prepared. Contamination during service will permit transfer of disease organisms directly to the consumer.

Serving and Eating Utensils

(A) Thorough washing and disinfection of utensils removes food particles to prevent growth of microorganisms and maintain aesthetics of foodservice. Utensils must be subjected to 170° F (77° C) for minimum of one half minute after cleaning to disinfect. Use of chemical germicides at room temperature requires ten minutes or more exposure to the chemical. Check concentration versus time.

(B) No dishes or utensils that are chipped, cracked, dented, or that contain crevices should be used. Such flaws provide hiding places for food particles and microorganisms. Heat and chemicals require more time to reach such hiding places than those times listed above, which apply to smooth surfaces.

Insects and Rodents

(A) All preparation and service areas should be free, and remain free of insect and rodent presence. (1) Automatically closing doors in double entranceways will help prevent entry of insects and rodents. (2) Buildings should be rodent-proofed. (3) Periodic closure to permit extermination service will aid in preventing infestation. (4) Thorough cleaning of all counters, tables, booths, and floors will remove food particles and trash to prevent attraction of insects or rodents.

Handling Dishes and Utensils

(A) The server's hands should never touch a surface that will come in direct contact with the mouth or food. Transfer of microorganisms from hands to dishes or food provides certain transmission to the consumer.

(B) Areas on dishes or utensils that touch counter or table tops should not be allowed to contact foods.

Foods

(A) Breads or pastries should always be handled with tongs or spatulas to avoid contamination from hands.

(B) Food, once served to a consumer, although not touched, should never be served again. This practice is specifically prohibited by most ordinances. The only exception is when foods are individually wrapped and remain unopened.

Buffet or Cafeteria Service

(A) Warming tables should *maintain* foods at or above 140° F (60° C) to prevent the growth of microorganisms in these foods.

(B) Cooling tables should *maintain* foods at or below 45° F (7° C) to prevent growth of microorganisms.

(C) Serving utensils should not be handled by customers. Service should be entirely by authorized workers employed by the establishment for that job.

(D) Servers should wear hair restraint, and should avoid putting hands to the hair while serving. This prevents contamination of food by loose hair, dust, or microorganisms, and reduces contamination of the hands.

(E) Transparent shields should be placed between the customers and the service tables and dishes. These shields (1) prevent respiratory contamination of foods by customers, and (2) prevent customers from handling and picking over foods. Maintenance of proper temperatures will help prevent growth of microorganisms; however, the warming table temperature will require several minutes to kill, and other temperatures will not kill.

Personnel

(A) Rigid compliance with local ordinances and health codes concerning the existence of known or suspected infectious disease in any foodservice personnel is essential to prevent spread of respiratory, contact, or foodborne infections.

Inspections

(A) Inspections by local health authorities should be encouraged. Many local health units have standardized check sheets for foodservice establishments. It would be well to obtain a copy of such check sheets and to use them in maintenance of the sanitary condition of the establishment.

(B) Records should be maintained of all inspections, and copies obtained of reports, deficiencies, corrections, and so on. Deficiencies should be corrected promptly and completely.

BIBLIOGRAPHY

AMERICAN PUBLIC HEALTH ASSOCIATION. 1985. *Standard methods for the examination of water and wastewater.* 16th ed. Washington, DC: Author in conjunction with American Water Works Association and Water Pollution Control Federation.

BENENSON, A. S., ed. 1985. *Control of communicable diseases in man.* 14th ed. Washington, DC: American Public Health Association.

JAWETZ, E., MELNICK, J. L., and ADELBERG, E. E. 1984. *Review of medical microbiology.* 16th ed. Los Altos, CA: Lange Medical Publications.

TEXAS DEPARTMENT OF HEALTH. 1984. *Rules on food service sanitation.* Austin, TX: Texas Department of Health, Division of Food and Drugs.

U.S. DEPARTMENT OF HEALTH, EDUCATION, AND WELFARE. 1976. *Food service sanitation manual.* DHEW Pub. no. (FDA) 78-2081. Washington, DC: U.S. Government Printing Office.

WOLFF, S. M., ed. 1985. *1986, The year book of infectious diseases.* Chicago: Year Book Medical Publishers, Inc.

5

Chemical Additives and Adulterants

The Food, Drug, and Cosmetic Act as enacted into law by the United States Congress in 1938, and as since amended, prohibits the adulteration of foods by any and all means. The scope of *adulteration* includes the addition of any substance or substances that are not a part of the natural foods. The law then permits "regulations fixing and establishing for any food . . . a reasonable definition and standard of identity. . . ." Obviously, no addition to foods is to be permitted that will be injurious to health. The "reasonable definition and standard of identity" phrase requires the identification of any deliberately added substance, and permits the setting of limits or tolerances for substances that may have been inadvertently added to the food product. The importance of chemical additives and contaminants has been increasingly emphasized in recent years as humans have gained knowledge of chemical effects on life processes.

For thousands of years, humans have added chemicals to foods. Smoke was probably the first food additive added during cooking, and was accidentally discovered to have preservative action. Probably not too much later, fermentation was discovered, and shortly, fermentation starters for breads and beers became additives. Certainly among the first additives to foods was salt, used ages ago not only for preservation of foods, but also to improve their flavor. In the Middle Ages, the search for sources of spices began a history of trade that continues to this day. In the earliest days, great importance was attached to such spices as pepper, nutmeg, cinnamon, and ginger for masking the flavor of slightly tainted or spoiled foods, and eventually these came to be valued for their additions of more desirable flavors to foods. Numerous other examples of the use of intentional or accidental food additives exist. Honey was among the first sweeteners to be used to change the flavors of food, and most likely the addition did not cause the great discussions we have heard in recent years about the safety of sweetening compounds.

Chemical substances that are not a natural component of the food may be classified as additives, adulterants, preservatives, or contaminants. The technical definition of a food additive of the Food Protection Committee of the Food and Nutrition Board, National Academy of Sciences is "a substance or mixture of substances other than a basic foodstuff which is present in a food as a result of any aspect of production, processing, storage or packaging." According to Schmidt (1974) the FDA recognizes thirty-two different categories of food additives that impart desirable characteristics to various foods. These categories are listed in Table 5.1. All of these substances are of concern to those involved in food sanitation from the standpoints of identity of foods, nutritive value, and potential effects on the health and well-being of the consumer. In recent years food additives have come to be of great interest to many persons because of adverse reactions to some chemicals by some individuals. In 1958, a food additives amendment was passed by the U.S. Congress which had four important aspects. First, the amendment defined food additives. Second, it authorized the FDA to license the use of food additives. Third, it permitted the proposed use of food additives if such can be proven safe when present at a certain level in foods. Perhaps most importantly, it states that if a proposed food additive is proven to be a carcinogen in animals or humans, its addition to food is then not permitted. This last provision has come to be known as the Delaney Clause or Delaney Amendment. Other provisions in this 1958 amendment provide regulations of pesticides, color additives, and new animal drugs in foods.

Foods are chemicals. The foods we consume daily begin as animal or plant materials and as such are composed primarily of proteins, fats, and carbohydrates with varying amounts of minerals, water, and other organic chemical molecules present. These other organic chemicals are classified as vitamins or growth factors in many cases. Also present in these natural foods are some amounts of chemicals that give color, flavor, and a particular texture to the food. In fact, foods are composed of specific combinations of most of the chemical elements that make up the matter of the earth. The particular combination and proportion of these elements in a food imparts the characteristics of that food that make it desirable to the consumer and that have nutritional value to the consumer.

Table 5.1.   Categories of Food Additives

Category	Common Uses
Anticaking	Added to crystalline foods (salts). Prevent lumping.
Antimicrobial agents	Prevent growth of microorganisms.
Antioxidants	Prevent rancidity or discoloration.
Colors	Impart, preserve, or enhance colors.
Curing & pickling agents	Impart flavor and color and increase shelf life (i.e., cured meats).
Dough strengtheners	Increase loaf volume; prevent loss of leavening gas.
Drying agents	Maintain low moisture. Prevent sogginess.
Emulsifiers	Preserve uniform dispersion.
Enzymes	Promote desirable chemical reactions.
Firming agents	Prevent collapse of tissues during processing (i.e., in pickles).
Flavor enhancers	Enhance and supplement tastes.
Flavoring agents	Impart change to desirable taste.
Flour-treating agents	Improve color and baking qualities of flour.
Formulation aids	Promote desired physical state or texture.
Fumigants	Volatiles to control or eliminate insects or pests.
Humectants	Promote moisture retention.
Leavening agents	Produce or increase carbon dioxide production.
Lubricants & release agents	Antistick agents.
Nonnutritive sweeteners	Less than 2% of the calories of sucrose.
Nutrient supplements	Improve nutritional quality.
Nutritive sweeteners	Greater than 2% of calories of sucrose. May give some physical texture.
Oxidizing & reducing agents	Produce more stable product. May add flavor.
pH Control agents	Maintain acidity or alkalinity.
Processing aids	Enhance eye-appeal. Include clarifying agents, clouding agents, flocculants, crystallization inhibitors.
Propellants, aerating agents, & gasses	Supply force to expel product from container.
Sequestrants	Effectively remove metals from chemical reactions and form more stable product.
Solvents & vehicles	Extract or dissolve another substance in processing.
Stabilizers & thickeners	Produce viscosity. Improve consistency. Impart body. Stabilize emulsions.
Surface-active agents	Modify surface tension properties of liquids.
Surface-finishing agents	Preserve & enhance surface appearance.
Synergists	React with other ingredients to produce total effect greater than the sum of the effects of individual ingredients.
Texturizers	Affect appearance or mouth-feel.

Source: *Federal Register*, 39:34173–34176, 1974.

INTENTIONAL FOOD ADDITIVES

To make foods more uniform in appearance, nutritional value, and quality, we have long accepted, or in fact demanded many treatment processes that alter the chemical content of the food. Many of these processes began as we recognized that there were inherent and unpredictable variations in the desirability, actual quality, and chemical composition of natural foods. In plant foods, such variations are in part due to environmental factors such as soil, moisture, temperature, and light, and in part due to the genetic makeup of the plant. In animal foods, the greatest effects on food composition and quality are likely to come from genetic factors; however, environmental effects are also frequently observed. To improve appeal, and to approach standardization of quality we have learned to apply specific processes to different foods. Grains are milled and blended to obtain flour of uniform baking characteristics, and nutrients removed by that process are returned to the flour as additives. Color is added to many foods to enhance attractiveness. Emulsifiers and thickeners are added to alter texture. Table 5.1 lists the many categories of food additives used today.

According to the FDA, four very common materials make up over ninety percent of the weight of all intentional food additives used in the United States: sucrose (sugar), salt, corn syrup, and dextrose. These and other additives were used (in 1970) in an amount of approximately 137 pounds per person per year. Most of the remaining 13.7 pounds per person can be found as starch, yellow mustard, sodium bicarbonate, yeasts, caramel, citric acid, carbon dioxide, black pepper, caseinates, phosphates, and glycerides for emulsification. Obviously then, if one assumes that there are around 2,000 additive materials in general use, the majority of them are going to be used in very minute amounts.

One may find vitamins, minerals, amino acids, or other organic growth factor supplements added to improve the food value of a basic or processed product. As long as these supplements are identified as to nature, source, and amount, and are among those accepted as adding nutritive value to the food, there is no general concern with their addition. The implication, however, that an additive has nutritive value when such has not been adequately demonstrated is not acceptable.

It is through practices such as these that we have gotten away from complete dependence upon natural foods, and have become increasingly concerned with the chemical content of foods, including the possibility of detriment to the nutritive value of some foods, and/or toxicity to the consumer. Although it is fully recognized that natural foods may contain small amounts of chemicals that may be harmful in large quantities, and some naturally occurring materials are poisonous to humans, the addition of harmful chemicals must be of concern to consumers as well as to producers of foods. In recent years the majority of confirmed foodborne disease caused by chemicals has been the result of consumption of seafoods, fish, or shellfish that are either toxic or have been inadequately preserved (Tables 5.2 and 5.3). Statistics also of concern, however, are those dealing with disease outbreaks caused by heavy metals, monosodium glutamate, mushrooms, sulfites, and miscellaneous chemical contamination because of the consistency with which these foodborne diseases occur. As in the case of microbial foodborne disease, it must be assumed that there are many more outbreaks and cases than are reported, and again the reports are only of confirmed instances, and do not include those presumptives that were not fully diagnosed.

Because of relatively low numbers of chemically caused foodborne disease outbreaks, there is no preponderance of evidence indicating that there is greater danger of food being contaminated in commercial establishments than in the home. However, in 1975 the only incidents of poisoning in the home involved the consumption of toxic mushrooms; chemical contaminations occurred in those places where food was likely to be handled by more people. In the outbreaks of poisoning from marine fishes, the source of mishandling, if indeed it did occur, could be much more difficult to locate, and the majority of these were unknown.

It has become acceptable to use chemical food additives when these

Table 5.2.   Confirmed Foodborne Disease Outbreaks and Cases Caused by Chemicals

Chemical	1973		1974		1975	
	Outbreaks	Cases	Outbreaks	Cases	Outbreaks	Cases
Heavy metals	0	0	4	28	4	50
Monosodium glutamate	2	6	2	4	3	9
Mushroom poisoning	9	41	6	9	5	5
*Marine fish poisoning	12	326	37	176	25	86
Paralytic shellfish poisoning	1	3	1	4	0	0
Neurotoxic shellfish poisoning	1	4	1	1	0	0
Miscellaneous chemicals	3	12	6	19	6	38

*Includes: Ciguatoxin, puffer fish tetrodotoxin and scombrotoxin.

Data compiled from CDC Foodborne and Waterborne Disease Outbreaks, Annual Summary 1975. Issued September 1976. U.S. Dept. of Health, Education and Welfare, Public Health Service Pub. # (CDC) 76-8155.

Table 5.3.   Nonbacterial Food Poisonings

Source	Disease Name	Latent Period	Symptoms
Wild mushrooms	Mushroom poisoning	½–2 hr.	Nausea, vomiting, diarrhea, cramps
Gray enamelware, acid foods	Antimony poisoning	Minutes–1 hr.	Vomiting, abdominal pain, diarrhea
Plated utensils, acid foods	Cadmium poisoning	¼–½ hr.	Nausea, vomiting, cramps diarrhea, shock
Copper pipes & utensils, acid foods	Copper poisoning	Minutes–hours	Nausea, vomiting, abdominal pain, diarrhea
Contaminated dry foods—Sodium Fluoride	Flouride poisoning	Minutes–2 hr.	Vomiting, diarrhea, abdominal pain, shock
Lead in utensils or as contamination, high-acid foods	Lead poisoning	½ hr.	Metallic taste, milky vomitus, black stools, shock
Tinned cans, high-acid foods	Tin poisoning	½–2 hr.	Nausea, vomiting, cramps diarrhea, headache
Galvanized utensils, high-acid foods	Zinc poisoning	Minutes–hours	Abdominal pain, dizziness nausea, vomiting
Cured meats or water	Nitrite Poisoning	1– hr.	Nausea, vomiting, cyanosis, weakness, headache
Several mushroom species	Cyclopeptide or Gyromitrin mushroom poisoning	6–24 hr.	Abdominal pain, bloating, vomiting, diarrhea, thirst, muscle cramps, jaundice, coma, death
Detergents, poorly washed bottles, beverages	Sodium hydroxide poisoning	Minutes	Burning lips, mouth, throat; vomiting; diarrhea
Some mushrooms	Iteolenic acid mushroom poisoning	½–1 hr.	Drowsiness, intoxication, muscle spasms, visual disturbances
Some mushrooms	Muscarine mushroom poisoning	¼–2 hr.	Sweating, salivation, low blood pressure, blurred vision, breathing irregularities
Insecticide-contaminated foods	Organophosphorous poisoning	Minutes–hours	Nausea, vomiting, abdominal cramps, diarrhea, headache, blurred vision, cyanosis, confusion

Source	Disease Name	Latent Period	Symptoms
Mussels & clams	Shellfish poisoning	Minutes–½ hr.	Tingling, burning lips, confusion, respiratory paralysis
Puffer fish	Tetrodon (puffer) poisoning	10 Min–3 hr.	Tingling, dizziness, twitching, paralysis, cyanosis, gastrointestinal disturbance
Tomatoes grafted to jimsonweed plant	Jimsonweed poisoning	Under 1 hr.	Thirst, photophobia, visual disturbance, coma, rapid heart rate
Root of water hemlock	Water hemlock poisoning	¼–1 hr.	Nausea, vomiting, stomach pain, convulsions, respiratory paralysis
Insecticide-contaminated foods	Chlorinated hydrocarbon poisoning	½–6 hr.	Nausea, vomiting, dizziness, muscle weakness, confusion
Ciguatera fish	Ciguatera poisoning	3–5 hr.	Tingling, numbness, dry mouth, gastrointestinal distress, watery diarrhea, paralysis
Grains, fish with mercury contamination	Mercury poisoning	1 wk. or longer	Numbness, weakness, paralysis, visual disturbance, coma, death
Cooking oils	Triorthocresyl phosphate poisoning	5–21 days	Gastrointestinal distress, leg pain
Tuna, mackerel, Pacific dolphin	Scombroid poisoning	Minutes–1 hr.	Headache, dizziness, nausea, vomiting, stomach pain
Chinese food (Monosodium glutamate)	Chinese restaurant syndrome	Minutes–1 hr.	Burning sensation upper body, dizziness, headache, nausea
Contaminated meats or other foods	Nicotinic acid poisoning	Minutes–1 hr.	Flushing, tingling, itching skin, abdominal pain
Organ meats of Arctic animals	Hypervitaminosis (Vitamin A)	1–6 hr.	Headache, gastrointestinal symptoms, dizziness, convulsions, collapse

additives constitute an advantage to the consumer. This advantage accrues when the additive (a) improves or preserves nutritional value, (b) improves keeping quality and makes the food more widely and readily available, or (c) enhances quality and consumer acceptance. The definition of a food additive as stated in the 1958 amendment to the Food, Drug, and Cosmetic Act of 1938 is

any substance the intended use of which results or may reasonably be expected to result, directly or indirectly, in its becoming a component or otherwise affecting the characteristics of any food (including any substance intended for use in producing, manufacturing, packing, processing, preparing, treating, packaging, transporting, or holding food; and including any source of radiation intended for any such use), if such substance is not generally recognized, among experts qualified by scientific training and experience to evaluate its safety, as having been adequately shown through scientific procedures (or, in the case of a substance used in food prior to January 1, 1958, through either scientific procedures or experience based on common use in food) to be safe under the conditions of its intended use.

When it is not in the best interest of the consumer, intentional food additives should not be used, and at all stages of food production, processing, preparation, and service, care should be taken to ensure that unintentional or accidental additives are not allowed to contaminate the food. It is not in the best interest of the consumer for additives to be used to disguise or hide poor quality, damage, or spoilage. It is not in the best interest of the consumer for food additives to be used in greater amount than necessary to achieve the desirable effect, or for the additives to be used instead of good processes, sanitary practices, and normal care.

The Baking Industry

The baking industry may be cited as an example where desirable use of intentional additives takes place. Yeasts growing in dough produce carbon dioxide from the carbohydrates present, and this gas, forming bubbles, causes the dough to rise, and after baking gives the bread a lighter, better texture. A number of chemicals can be used to substitute for the action of the yeast; for example, baking powder, a mixture of chemicals that act as acids in the presence of moisture and heat. This chemical additive produces an advantage to the consumer in lighter, better textured bakery products in which flavor and texture may be better controlled. The additive is not harmful, and in fact, chemical reaction in the food results in very little of the original additive chemical remaining in the finished product.

Natural wheat flour is not the white product that we are most used to seeing. Components of the wheat make this flour beige or light yellow in its natural state. Bakers recognized long ago that upon aging, this wheat flour became white, and that the baking qualities were improved in white flour. Chemists found that the addition of certain oxidizing agents would speed up the whitening and aging processes, and would not interfere with the good baking characteristics of the flour. The use of these additives reduces the previously required storage time, avoids the danger of spoilage or contamination during storage, and thereby reduces the cost to the consumer while protecting the quality of the product.

The baking industry in practice illustrates many of the features of the practice of use of GRAS substances. These are chemical additives that have been used for a number of years, and through use, have gradually come to be accepted as safe in the manner in which they are generally used. When the Food Drug and Cosmetic Act was amended in 1958, there was established a requirement for premarket safety approval of food additives, but there was an exemption for ingredients of food in common use at that time—these have since become known as GRAS substances. They include those additives or ingredients that are "generally recognized among scientific experts qualified by scientific training and experience to evaluate the safety of chemicals to be safe under the conditions intended for use." This amendment defined food additives (as above), authorized the Food and Drug Administration to license the use of food additives; permitted the use of food additives if they are proven safe when present at certain specified levels in foods, and forbade the use of a food additive if that additive is a proven carcinogen in animals or humans. The latter clause has come to be known as the Delaney Clause, and continues to cause some disagreement and argument among some experts. This amendment has become a most important aspect of the Food and Drug Act because of the continuing increase in use of food additives, and the application therein of both the GRAS definition, and the Delaney Clause. Other provisions of the 1958 amendment regulate the presence of pesticides, color additives in foods, and new animal drugs.

Since the establishment of the GRAS category of additives by this amendment, the FDA has developed many lists of substances that may be considered as GRAS. These lists are frequently reviewed, modified, and published, although the FDA has, itself, stated that it is impracticable to list all substances that are generally recognized as safe for their intended use, and has stated specifically that such additives as salt, pepper, vinegar, and sugar are safe for their intended use, and these condiments are not reviewed and retested. Published in 1979,

Part 182 of the Code of Federal Regulations lists those substances that were considered to be GRAS when used in accord with good manufacturing practices.

Other Uses

A number of chemicals may be added to foods as emulsifiers and thickening agents. Many materials used for these purposes are also fillers of low caloric or nutritive value. While these additives may improve the appearance and texture of foods such as frozen desserts, candies, chocolates, and ice cream, they do not add nutritive value to any extent, and may make the foods nutritionally expensive. They may be desirable in some diets for weight control, in addition to the aesthetic improvement of the food's appearance.

Food flavors and colors are enhanced by a wide variety of both natural and synthetic chemicals. These materials present a constantly changing array as each is studied more thoroughly for its indirect effects as well as its intended effects. Monosodium glutamate (MSG) is such a chemical which has been used for many years to intensify flavors of certain foods and to enhance other natural flavors. It is now recognized, however, that this chemical is responsible for a type of foodborne disorder in some persons known as Chinese restaurant syndrome. In susceptible individuals who have consumed this chemical, headache, weakness, malaise, and muscular tingling results within a half hour after eating. These symptoms may closely resemble those of some types of mushroom poisoning. The recognition of this chemical effect is indicative of the need for careful study and long-term observation of food additives.

Another food additive, used in several forms for preserving the freshness and fresh appearance of many types of foods, is commonly termed *sulfites*. Sulfiting agents include sulfur dioxide and other forms of inorganic sulfites that release sulfur dioxide when used as food ingredients. These can be recognized on food labels as sulfur dioxide, potassium bisulfite, potassium metabisulfite, sodium bisulfite, sodium metabisulfite, or sodium sulfite. These are very often included in such foods as avocado preparations, dried codfish, fruit, gelatins, potatoes, salads, sauces, gravies, sauerkraut, cole slaw, shellfish, soups, fresh vegetables, and vinegars; and in drinks such as beer, cider, wine and wine coolers, and fruit juices. Some of the reasons these chemicals are used include the fact that sulfur dioxide will bleach food starches; sodium metasulfite and sodium sulfite prevent rust and scaling in boiler water used in making steam; sodium sulfite and sodium metasulfite are used in producing cellophane for food packaging; and potassium metabisulfite, sodium bisulfite, sodium metabisulfite, and sulfur dioxide are used in wine making as sterilizing agents.

The Food and Drug Administration is taking steps that will help consumers be more aware of those products that contain sulfites, and to reduce exposure of the general population to foods that have been incriminated as being responsible for adverse reactions in the past. The FDA received about 850 complaints between January 1 and mid-September of 1985 of adverse reactions to sulfites. Among these 850 cases, 20 deaths occurred. Most of these complaints involved foods or beverages; however, some also involved drugs. Many complaints occur after consumption of raw fruits and vegetables in restaurants, a few deal with fresh produce bought in retail grocery stores, and a number involve beer and wine. The FDA continues to receive approximately two complaints a day blaming the presence of sulfites in food or drink as the cause of adverse reactions. The reactions reported frequently include symptoms of wheezing, vomiting, nausea, diarrhea, unconsciousness, abdominal pain, cramps, and hives. Early in 1985, the FDA ruled that shrimp that contained residues of sulfites above 100 parts per million would be considered to be adulterated. An FDA ban on the use of sulfites in foods would not completely solve all the problems of adverse reactions to these chemicals because sulfur dioxide is used as a fungicide in the growing of grapes, and the use of sulfites in wine and beer making does not fall in the jurisdiction of the FDA. The FDA does, however, continue to attempt to make jurisdictional rulings to reduce the exposure of many persons to sulfites through food consumption.

Spices and chemicals to add or alter color and flavor of foods have been used since early in human history. Properly used after adequate study, these chemicals have added much to the enjoyment of food throughout the ages. They have also been used to hide poor quality and even spoilage, and when used in this way violate all standards of good practice, sanitation, and honesty.

In the processed foods of today, it is not uncommon that the processes used have removed or destroyed some desirable components of the food. In some cases, food quality may be improved simply by addition of some chemicals. These situations have been recognized particularly in regard to some of the vitamins and minerals needed in the human diet. Among foods recognized to be improved by such additions are breakfast cereals, breads, milk, margarine, cheeses, pastas, and so on. For addition of vitamins and minerals to these products, the FDA has set definitions and standards of identity of materials used for enrichment.

Although processing or cooking destroys most of the microorganisms in foods, the process may also expose some foods to additional contamination by microorganisms or their spores. For this reason, certain chemicals, after careful consideration and study, may be added to some foods as preservatives to prevent microbial growth and certain chemical reactions. The use of sulfites, previously cited, is an example of this. Other foods to which preservatives are added are baked goods, cheeses, syrups, margarine, mayonnaise, shortening, bacon, chips, certain salads, fruits, and so on. In these foods, the chemicals prevent potentially toxic growth of microorganisms, or prevent flavor change or preserve or improve texture and flavor. Of the numbers of chemicals used in this way, some have become suspect as not being completely harmless to the consumer, such as monosodium glutamate and sulfites, as previously mentioned. Another such chemical is nitrite, which is added to bacon and some other meats to prevent growth of the bacterium, *Clostridium botulinum,* the producer of a dangerous and highly lethal toxin under the right conditions of growth. Nitrite is very effective in preventing *C. botulinum* growth when used properly; however, there is some evidence that it may cause the production of cancer in experimental animals when it is used in high doses—much higher than that needed for the preservative action in meats. In the case of this chemical, as in many others, it should be remembered that nitrites are components of the nitrogen cycle that operates naturally in surface waters and soils, and is essential for all life. Thus, the true danger to the consumer from possible botulism must be weighed against the much more remote chance of production of cancer when small amounts are used to preserve certain foods. Chemicals such as this are subjected to intensive study by the food industries, and private and governmental agencies, and frequently new regulatory directives result from these studies.

Certain chemicals that are safe, but that provide no direct benefit to the consumer may be added to foods. Such additives are exemplified in the addition of water to meats by injection. This practice may improve cured ham, for example, by improving texture and palatability. However, because no food value is added, and this ingredient does add weight, the addition must be specified on the product label. If such addition is not specified, the food may justly be considered to be adulterated.

Although the practices of addition of spices, colors, preservatives, and other chemicals to foods have been of long standing, the continued studies and newly discovered information about these materials make it necessary that workers in food industries continue to update their knowledge of these practices by keeping abreast of research publications and governmental directives as this field continues to develop.

CONTAMINANTS

Those chemicals that may be of greatest concern to sanitation are included in the group classed as contaminants. It is in this class that the greatest increase in chemical varieties has occurred in recent years. Chemicals that may be present as contaminants include fertilizers, insecticides, fungicides, herbicides, defoliants, rodenticides, and antibiotics. The use of such chemicals has resulted in great strides in the control of plant, animal, and human disease; in ridding farmlands of useless or harmful plants, thus permitting greater yields of foods from these lands; in controlling animal diseases and stimulating animal growth resulting in greater food yields; and in stimulating plant growth rates and total yields. It is therefore easy to understand that the intended purpose for each of these chemicals has been successfully attained, and that these purposes and these uses have been beneficial to the human in the majority of cases. We are frequently reminded that fewer and fewer farmers produce more and more foods on less and less land. This has been accomplished in the United States by the development and careful use of many kinds of technology, including chemical control of pests of all varieties. The success of this control has, in a very large measure, been responsible for the success we observe in improving human health and well-being.

The opposite side of the coin is that chemicals used to control pests of all types do have some adverse effects on humans and on beneficial life if these chemicals are used indiscriminately or unwisely, and if these chemicals by accident contaminate foods to be consumed by the human. We now hear much of banning the use of various pesticides because of harmful effects. That these harmful effects have occurred, do occur, and will occur, is all too true. That these harmful effects are most often due to unwise, unwarranted, or improper use of the chemicals is also true. Most insecticides are poisonous to the human if taken internally, or even if contacted externally in concentrations that are too large. Most of these pesticides will kill harmless or even beneficial organisms of some varieties in addition to killing those insects for which the chemical is used. Human deaths have resulted from contact with excessive amounts of these chemicals. As a result of these consequences of misuse, the cry is raised by many that these chemicals should be banned for use in all circumstances.

A like case could be made for antibiotics, the wonder drugs of the twentieth century which have done so much to alleviate human disease and suffering and to prevent human death. Many antibiotics taken internally in concentrations exceeding the proper dose are poisonous to humans. These antibiotics kill beneficial as well as disease-producing microorganisms. Human deaths have resulted from improper or accidental overdoses or uses of antibiotics. Use of antibiotics has saved countless lives and results in more persons being alive, thus adding to the imbalance of nature in the human population explosion. Yet, will the human who demands the banning of insecticides (which may control disease-carrying mosquitos) deny himself, herself, or a sick child the benefits of proper antibiotic usage because of similar potential dangers? It is better to advocate the *proper* use of chemical agents of pest control of all kinds and thus to minimize the dangers of use while gaining the many benefits of properly regulated usage.

Much of the potential danger resulting from the use of chemical agents of control for microbial, plant, and animal life lies in the possible contamination of human foods with these materials. In many instances this contamination occurs because residues of the chemical remain after use aimed at control of pests. Such residual pesticide on foods is undeniably undesirable, and, even in cases of proper usage, may be unavoidable. A vegetable crop nearing maturity for best harvest becomes most attractive to destructive insects. The crop must be treated with pesticide or be destroyed. In such treatment, procedures are recommended that provide sufficient time for the insecticide to be removed or degraded before harvest. It is inevitable, however, that in a few cases, some insecticide may remain. Because extensive testing has shown that many insecticides are not harmful in very, very small quantities, and are readily degraded and removed from animal bodies, the Food and Drug Administration has set certain tolerances or limits of concentration for certain pesticides that will be allowed in certain foods. Such tolerances are set only for raw foods, and if a specific tolerance has not been set, then any amount of that pesticide in a food is prohibited. If a pesticide has been shown to be definitely and completely harmless, then it may be granted exemption and may be allowed to be present in minute amounts in raw foods. Such exemptions are rare and it is generally true that if such chemicals are present, the manner of use of the food is such that the chemical will be removed or degraded before consumption. Pesticides present in raw foods are usually present on the surface and not in the internal portion of the food. If this food is one that must be peeled before use, then the pesticide that is present will be removed. Tolerances are more applicable to foods not peeled or shelled before preparation or consumption. The use of chemicals for control of pests, weeds, and diseases in plants and animals makes possible the bountiful supply in this country. Recognizing the dangers as well as the economic considerations and benefits involved in use of chemicals, constant research efforts are being made to devise or discover new and safer methods of control of the environment for food production.

For example, while the use of insecticides has prevented loss or damage to food crops amounting to millions of dollars each year, it is possible that some insects may be controlled as well or better and more safely by a natural enemy such as an insect predator, and the balance of nature may be better maintained. Until such control is established, however, insecticides are tested repeatedly during development for both acute and chronic toxicity to animal species. It is rare that a pesticide will be approved for use with less than three years of testing by the manufacturer. Food processors have in many cases established their own laboratories for research aimed at detecting pesticide residues and at removing these residues from foods. In addition, both federal and state agencies are assigned responsibilities for continuous testing of products to assure that human health is protected. Generally, the Health Departments of the various states are the responsible agencies at that level, while federal responsibility may be shared by or divided among several agencies of the Department of Health and Human Services (including FDA and CDC), the Department of Agriculture, and the Environmental Protection Agency.

Plant and animal growth regulators are used in a variety of ways to improve food production. Such chemicals may be used to inhibit the growth of weeds and grasses in crop fields; to promote bloom and bearing times at earlier dates in fruits; to assist in thinning fruits on bearing trees; and to speed up the bearing age of young plants. Animal growth stimulation may result in earlier maturity of the animal, or in greater size of the animal at time for slaughter.

Fertilizers, used properly, greatly increase the yields of plant crops, and result in maximum production while assisting in maintenance of soil fertility. Fungicides, bactericides, antibiotics, and so on, used to treat or prevent diseases in plants and animals, prevent loss or damage to food crops, help to stimulate animal growth by inhibiting parasitic growth, and help to prevent transmission of animal diseases to humans.

With the uncounted benefits of each of these kinds of synthetic and natural chemical usages must come the recognition that these chemicals used artificially and in unnatural locations may well upset the balance of nature; may result in pollution or chemical contamination of water supplies; and may result in some residuals in foods. Recognition of the problems must simply be followed by regulation and use of such controls as will produce maximum benefit and minimum hazards to humans and their environment. Obviously, the ingenuity of the human who is capable of producing such agents is also sufficient for permitting the development of effective measures for controlling both benefits and hazards of use of such chemicals. Some difficult lessons have been learned in chemical control of human diseases; and these lessons may well be applied to other types of chemical controls to produce the desired effects.

In addition to the contaminants that may be present as residuals, other types of foreign materials may enter foods as accidental contaminants. This includes all those chemicals that might be used in food preparation areas either in the home, foodservice establishment, or processing plant such as soaps, detergents, disinfectants, cleansers, polishes, household insecticides, rodenticides, or air fresheners. Such substances may contaminate utensils or surfaces, which then serve as the vehicle for transfer of the contaminants to food with which they come into contact. Who has not, at some time, looked forward to that fresh cup of coffee only to have it taste distinctly of dishwashing soap or detergent that has not been thoroughly rinsed away after the last cleaning? Mechanical dishwashers are usually very efficient, with jet sprays and mechanical propulsion of rinse water so that all portions of any article subjected to such cleaning is properly rinsed. The efficiency, however, depends on the good judgment and proper operation of the dishwasher by the person who is loading and operating it.

Disinfectants, cleansers, and polishes may be used on installed appliances, work surfaces, storage areas, and refrigeration spaces at frequent intervals between periods of food preparation. The frequent use of these chemicals in goodly concentration is essential for keeping any food work area and preparation space in attractive and sanitary condition. On some surfaces and in some machinery, disinfectants should be used in such a manner as to leave a residual for continuing action. This residual should not be left if the surface, appliance, or instrument is to come into direct contact with food in the process of preparation or serving. For surfaces and appliances of this type, the chemical in use must be completely removed and the surface left completely free of any material that can be transferred to the food.

In a food preparation area, insecticides and rodenticides should be used only when food is not being prepared, exposed, or handled. Contamination may occur if these chemicals are applied as sprays or vapors, as are air fresheners or deodorizers. A good method of preventing such an accident is the use of a paint or solid chemical pesticide. The effectiveness of the chemical should also be considered, as should the odor produced and the aesthetics of the application. In the application of pesticides, it is essential that materials not be used that will act in such a way that dying pests are able to reach foods and contaminate them. Adequate security in food storage areas will be sufficient, along with regular thorough housekeeping to prevent such occurrence.

Other accidental contaminants are likely to be particulate rather than soluble chemicals, and are easily controlled by housekeeping methods. Ordinary care, attention to details, and personal hygiene practices are sufficient to prevent contamination by bits of chipped glass or metal; hair, dust, or airborne particles of other kinds; or any foreign materials. Invariably such contaminations are the result of carelessness or sloppy personal habits of persons working in the area.

Controlling the presence of extraneous chemicals in foods is not an impossible task in spite of the prevalence and increasing use of chemicals for diverse reasons in our total environment. Before such chemicals can be sold commercially, directions for use and warnings of hazards must be explicitly stated on labels. The best control measure to prevent the contamination of, or residual in food of any undesired material is simply that the person applying chemicals *read labels completely, then apply the agent exactly according to every detail of directions*; and that persons working in food areas work as if their own health depended on sanitation practices and personal habits used. Details for use of chemicals are likely to include (a) proper concentration for use for different purposes, (b) precautions to be observed in mixing, (c) proper timing for use, (d) proper locations for use, and (e) safe methods of application. Any chemical licensed for general use in the environment can be made safe if the user understands and follows such directions. Dangers to individuals, and to the environment result when chemicals are used for the wrong purpose, in concentrations larger than needed or recommended, and applied by methods that do not permit proper control. Many chemicals are poisonous, and are so labeled, but they can poison only with the help of the human who uses them incorrectly.

The great benefits to humans of proper chemical use to control biologic environments may include aesthetic benefits occurring from such use. How much more desirable is a restaurant free of insects than one in which insect populations can be observed? How much more delicious the apple or peach without a blemish than ones that have been stung, scarred, or marred by insect or microbial attack? Proper chemical control of the environment for food production, preparation, service, and consumption has not only made the food safer from a health standpoint, but has made it more attractive, appealing, and nutritious. Regulated, careful, and properly timed use of chemical agents will retain these benefits for the human.

BIBLIOGRAPHY

CONNING, D. M., and LANSDOWN, A. B. G.   1983.   *Toxic hazards in foods.* New York: Raven Press.

HUI, Y. H.   1979.   *U. S. Food laws, regulations and standards.* New York: John Wiley and Sons.

LECOS, C.   1986.   Reacting to sulfites. *FDA Consumer* 19:17–20.

NATIONAL ACADEMY OF SCIENCES.   1973.   *The use of chemicals in food production, processing, storage, and distribution.* Committee on Food Protection, Food and Nutrition Board, Division of Biology and Agriculture, National Research Council. National Academy of Sciences. Washington, DC: Author.

PETERSON, M. S., and JOHNSON, A. H.   1978.   *Encyclopedia of food science.* Westport, CT: AVI Publishing Co., Inc.

ROBERTS, H. R., ed.   1981.   *Food safety.* New York: John Wiley and Sons.

SCHMIDT, A. M.   1974.   Designation of food categories and food ingredient functions. *Federal Register* 39:34173–34176.

6

Water Use in Food Industries

WATER SOURCES

Sanskrit and Egyptian inscriptions indicate that there was some knowledge of water treatment, particularly filtering and boiling, as early as the fifteenth century B.C. Hippocrates (460–354 B.C.) recommended that rainwater should be boiled and strained. Public water supplies were developed to some extent in these ancient times, and have become more and more important as human populations have become more urbanized.

All life processes require a constant source of water, and living organisms receive this supply of water by way of the natural recycling of the water supply of the earth in its three locations: air, land surface, and underground. Recycling of any materials generally implies reuse, and this is most certainly the case with water. It must be reused with considerable frequency if the living forms on the earth are to receive an adequate supply. Because the water we use is being reused, it is essential that we examine our current water source to determine whether it was made harmful by conditions existing in its prior use. Of the three sources of water for human use, groundwater located in deep reservoirs under the surface of the earth is the most nearly isolated from humans and is, therefore, least likely to be contaminated or polluted by humans and their activities. Unfortunately, groundwater is not always geographically located in sufficient quantities in areas where the needs are greatest, and when it is located near the place of need, it too often has also been exposed to some human activity, either through deliberate modification of the area by humans (i.e., landfill activities for solid waste disposal), or through some fault or defect in the structure of the surface of the land in the area so that the natural function of filtration or percolation through soil and rock formations does not occur. Groundwater volumes near large urban areas are frequently not sufficient for supply of the needs of those areas, and when the volumes are great there is some detriment that results if large amounts of groundwater are removed (i.e., subsidence).

Atmospheric water is a potential source for human water supply only as it is recycled to the land surface. In the process of this recycling, the air is washed by the falling water, and depending upon the geographic location, rather heavy pollution or contamination may be present in the water by the time it reaches the earth's surface. Currently, we depend upon the natural cyclic processes to move water from its atmospheric location back to the surface of the earth, and these natural processes do not always occur with the regularity that would be most desirable from a supply standpoint.

Surface water makes up a majority of water supply for human use. The major portion of the surface of the earth is covered by water in the oceans, lakes, rivers, ponds, and reservoirs. By building dams and creating reservoirs to retain water and prevent its rapid return to the saltwater of the ocean, we have increased the surface area covered by water, and, in effect, have changed the location, and chemical and biological content of a major portion of the available water supply of the earth. Because these surface waters are readily available to humans, in addition to convenience for use as a source of water, they are also most convenient as a potential means of removal of waste materials discarded by humans. The need for large drainage areas to supply and maintain surface-water sources automatically subjects these surface waters to contamination by way of run-off from urban and industrial areas. These run-offs are sources for much of the non-point source pollution of water supplies, and may be very important in consideration of the chemical content that may affect water quality for many activities. Many large cities in the United States take their water supplies for treatment from rivers or reservoirs that receive run-off from such large basins, and are in some danger of being affected by various forms of chemical or biological pollution from unknown sources. Examples of such cities would include New Orleans, Philadelphia, and Cincinnati. On the other hand, Boston, New York, and San Francisco are examples of cities that have accessed sources upstream which are less likely to be polluted for their water supplies.

Water may be polluted by the intentional or incidental addition of either chemicals or living organisms. Although many harmless materials may be present in water, we do not consider those as pollutants. A pollutant is a material, either living or non-living, that may be detrimental to the health or well-being of natural aquatic plant and/or animal populations, or to humans. At the present time, very little surface water remains in the United States that is fit for human consumption before treatment, and even in those areas where surface water appears to be fit, it is dangerous to trust such appearance because of the potential for presence of either chemical or biological pollutants even in the areas most remote from human habitations.

WATER POLLUTION

Water may be polluted by the addition of chemicals and/or living organisms that make it detrimental to the health and well-being of naturally aquatic plant and animal populations, or to humans. Water was suspected of being a source of disease long before waterborne disease transmission was proven just over 100 years ago. Since that time, humans have learned to prevent disease transmission by treatment of the water before consumption. More recently, chemical pollution of water has become of greater concern as people have carelessly dumped more waste materials onto the earth's surface and into surface waters. Chemical pollution affects not only the chemical content, but the biological content of the water as well. As chemicals are added indiscriminately to water, some will stimulate the growth of certain aquatic organisms while others will inhibit the growth of some living organisms. The net result is an upset of the natural aquatic ecological balance. This change may possibly result in (a) actual toxicity to the human consumer, either chemical or biological; (b) production of additional chemicals causing taste or odor problems that make the water undesirable; or (c) massive growths of certain living organisms that will literally fill the lake or reservoir with solid materials, thus rendering it useless as a water source after a period of time. In some instances, that period has been as little as forty years.

Because of the probability of pollution to some degree, of surface and subsurface (ground) water, Congress enacted the Safe Drinking Water Act of 1974 (Public Law 93-523), and it has since been amended, to assure the supply of safe, potable water to all U.S. citizens when that water is supplied by any governmental unit or commercial establishment. Under that law, the National Academy of Sciences was charged with studying potential pollutants and their possible effects on humans, and to report those studies to the Congress within a specified time. Those studies are now embodied in a series of five volumes entitled *Drinking Water and Health.* These studies describe in considerable detail the possible consequences of consuming water polluted by the presence of numerous inorganic and organic chemicals, and/or biologic entities. The Safe Drinking Water Act has as its objective the treatment of water supplies from all sources in such a manner that pollutants will either be made harmless, or will be removed.

Biological Pollution

Water pollution increases the potential hazards of using water. Water treatment processes are designed to remove biological hazards, and are extremely effective in reaching this objective. However, these processes must constantly be changed to compensate for the amount of chemical contamination, because much of the contamination will interfere with treatment processes aimed at removing biological hazards.

Untreated water consumed directly may contain a variety of pathogens and will thus serve as a vehicle for infectious disease transmission. Waterborne diseases of this type frequently occur as epidemics involving large numbers of persons, and waterborne diseases do not always require consumption for transmission to occur. For example, industrial manufacturing plants have had outbreaks of diseases similar to Legionnaires' disease when wash waters were used in certain plant operations, although no person actually consumed the waters that contributed to the disease outbreak. Among diseases that have been proven to be transmitted by water are cholera, typhoid fever, salmonellosis, bacterial dysentery, amoebic dysentery, and various types of gastroenteritis. Diseases that can be transmitted by contaminated water are discussed in more detail in *Procedures to Investigate Waterborne Illness* (1979). Untreated water used in cleaning vegetables and fruits that are consumed without cooking also will serve for transmission. Standard water treatment practices adjusted to the specific needs of a situation are adequate for prevention of transmission of these diseases, and have been successfully used for many years. The greatest flaw in these practices occurs in failure to adjust to special needs. An example is seen in the use of water to clean nuts, such as pecans, before they are processed. The pecan shell, during harvest, may accumulate filth or dirt that is composed largely of organic matter, and that may contain disease-producing bacteria. If these shells are washed in potable water, the chlorine present in this water is rapidly tied up or neutralized by the extraneous organic matter found on the shell, and disease-producing bacteria are not killed. If the shells are cracked, these living bacteria are able to enter and contaminate the nut meats, which are likely to be consumed without cooking. Depending upon the amount of extraneous organic material present, some 40 to 200 mg/l chlorine may be needed to provide a margin of safety.

Yet, despite the application of disinfectants to water and the use of advanced technological treatment processes for the physical removal of pollutants, it appears that from about 1951 through 1980, the average annual number of waterborne disease incidents has been on an increase. Table 6.1 presents reported outbreaks and number of cases for some states for the years 1960, 1975, and 1980. The data presented also include a differentiation, when available, between the largest incidence of cases of biological versus chemical causes for the outbreaks. In examining data such as these, it should be kept in mind that over the years the states have improved their reporting systems and public awareness of waterborne diseases has increased significantly. With this taken into account, it may very well be that more outbreaks are being recognized and reported, although there are no more outbreaks occurring. Some estimates of the ratio of outbreaks reported to the total, however, range as low as twenty percent. So there remains considerable room for improvement.

Table 6.1. Number of Waterborne Disease Outbreaks (O) and Cases (C)

State	1960		1975		1984	
	O	C	O	C	O	C
Arkansas	—	—	2	523	2	126
California	1	28	3	999	1	1
Colorado	—	—	—	—	3	463
Connecticut	1	23	—	—	—	—
Idaho	—	—	1	9	1	6
Illinois	1	2	—	—	—	—
Indiana	—	—	1	1,400	—	—
Louisiana	—	—	1	26	—	—
Massachusetts	—	—	1	17	1	7
Minnesota	—	—	1	136	1	9
Missouri	—	—	—	—	3	113
Montana	—	—	1	56	—	—
New Jersey	1	63	3	744	—	—
New York	4	1,511	—	—	1	4
North Carolina	—	—	—	—	1	1
Ohio	1	98	1	140	—	—
Oregon	1	6	1	1,007	3	84
Pennsylvania	—	—	4	5,225	5	456
South Carolina	—	—	1	7	—	—
Tennessee	—	—	1	40	—	—
Texas	—	—	—	—	2	368
Virginia	—	—	—	—	1	28
West Virginia	1	53	—	—	—	—
Wisconsin	—	—	—	—	1	89
Totals	11	1,784	23	10,329	26	1,755

Source: "Water-related Disease Outbreaks; Surveillance, CDC, Annual Summary, 1984, and Foodborne & Waterborne Disease Outbreaks, Annual Summary, 1975.
    Some cases were chemically related in Arkansas, Colorado, Florida, Idaho, Illinois, Louisiana, Massachusetts, Montana, New Hampshire, New Jersey, New York, North Carolina, Ohio, Oregon, Pennsylvania, South Carolina, Tennessee, Vermont, Washington and Wisconsin in 1974 and 1975. Total chemical in 1974 = 269; in 1975 = 37.

Chemical Pollution

Current technology is producing an increasing number of chemicals that, although useful, may eventually pollute water sources. This production also results in an increasing variety of waste chemicals, many of which may be discarded directly or indirectly into water sources. Some of these chemical pollutants are rapidly modified or destroyed by biological activities; others are left unaltered for years, and some have a direct, cumulative, and long-lasting effect on either the water or in some cases on potential food sources in the water such as fish or shellfish.

The types of chemicals that are of greatest concern to sanitation are those that may be residual and that are directly toxic to humans or to aquatic life. Examples are insecticides, herbicides, organic fertilizers, petroleum compounds, mercury compounds, arsenic compounds, and dyes. Standard water treatment processes are not always totally effective in removing these chemical hazards and the use or consumption of water polluted by these materials will serve to transfer toxicity to food products for human consumption, or directly to the human. Governmental regulations set limits or tolerances for many of these chemicals, and the presence of many others is prohibited altogether in foods.

Some chemical water pollutants are not harmful in very small concentrations, but when consumed by humans or animals are not readily metabolizable and can accumulate to a concentration that is toxic. Such accumulation of chemicals is most readily observed in aquatic life. Fish are frequently killed; or though not killed, accumulate sufficient quantities of the chemicals that when consumed, may be toxic to the consumer. Chemical analyses of fish, both from freshwater and marine sources, have demonstrated considerable accumulations of potentially toxic chemical pollutants in several geographic locations.

Recognition of the chemical hazards found in drinking water can be appreciated when one reviews the maximum contaminant levels (MCL) proposed by the U.S. Environmental Protection Agency when it proposed the National Interim Primary Drinking Water Regulations (J.A.W.W.A., 1976). Limits of heavy metals and their MCL's (mg/l) included arsenic (0.05), barium (1.0), cadmium (0.01), chromium (0.05), lead (0.05), mercury (0.002), selenium (0.01), and silver (0.05). The nitrate level remained at 10 mg/l (as N); formerly it was 45 mg/l (as NO_3). Chlorinated hydrocarbons, on the other hand, have MCL's at generally lower levels: endrin (0.0002), lindane (0.004), methoxychlor (0.1), and toxaphene (0.005). These are but a few examples of potentially hazardous chemicals in potable water supplies.

Food sanitation must be concerned with the state of chemical pollution of water sources (a) for direct consumption of the water, (b) for consumption by aquatic food sources, and (c) for use of the water in preparation of all foods. As the level of chemical pollution increases, food industries must either obtain water sources for operations that have minimum chemical pollutants, or seek methods for water treatment that will remove or neutralize any pollution that is present.

Treatment of water by standard methods generally results in a clear product that is not unappealing. Such treatment, however, will not remove many tastes and odors that aesthetically reduce the water quality. Undesirable tastes and odors of water are often the result of biological activities that produce new chemical compounds. Such chemicals are often quite stable and when such water is used in food preparation, the taste or odor is simply transferred to the food, and in some cases appears to be actually concentrated or accentuated in the foods.

Water sources for food industries should be selected for the absence of such chemical contaminants. Many waters, unless repeatedly distilled and treated, will contain dissolved chemicals that affect the taste. Perhaps everyone has noted some variation in the taste of soft drinks bottled in different geographical locations using different water sources. Such variances are the result of chemical content of the water, rather than the use of different recipes for manufacturing the product. When natural chemical contents of water can produce such noticeable results, it is obvious that chemical pollutants in water sources can result in completely undesirable or unpalatable food products.

WATER TREATMENT

Water treatment practices in the United States have been developed and regularly practiced and are overseen by governmental agencies. The basic methodology has proven effective for over seventy years. This standardization and regulation has resulted in safe, potable water supplies that are accepted as a matter of routine expectation. The operation of water treatment facilities generally is practiced in the United States as a function of municipal governments or water districts, which are governed by state regulations. The efficiency of this mode of operation is attested to by the decline of disease epidemics that can be traced to waterborne infectious organisms. Methods used for successful purification include several steps: sedimentation, pretreatment, coagulation, softening, filtration, and disinfection.

Sedimentation may be used at two stages of water treatment. Many water treatment facilities include a large basin or lake adjacent to the plant where water is held in large quantities before being taken into the plant. Such reservoirs are open, and are generally in themselves subject to contamination by groundwater run-off, and collection of precipitation that has removed air pollutants while falling. Such basins are, however, protected from waste pollution and, to a large extent, from animal pollution. The volume of these reservoirs provides an essentially stationary condition that permits larger particles of solid materials and plant growth to settle out of suspension. Settling of these masses will also trap and remove many smaller particles, including bacterial and fungal cells. The location of such a basin adjacent to the water treatment plant permits intake of water into the plant in such a way that sediments will less likely be resuspended and drawn into the treatment facilities, eliminating the necessity for further removal of this debris.

Water treatment plants that take in highly turbid river water usually have bar screens to separate the debris over 0.5 inch, followed by grit chambers (or primary sedimentation vessels) for sand removal. It is at this point that pretreatment is applied to the water if it is needed. If tastes and odors are present, powdered, slurried, activated carbon and/or potassium permanganate are added. Some detention time should follow this treatment before coagulation. Also, many plants chlorinate the raw incoming water. The addition of chlorine prior to coagulation and softening reduces the bacterial load and enhances coagulation and softening processes in many instances.

At this point it is advantageous to recognize the fact that some tastes and odors are caused by a group of organisms called the actinomycetes. These organisms and, on occasion, some blue-green algae, contribute chemical compounds to surface waters during certain times of the year. In the southwestern United States, the occurrence is usually from June to late September. Tastes and odors emitted from these organisms follow in succession. The order of occurrence, from the earliest phase imparted to the water to the last, is (1) grassy, (2) fishy, (3) woody, (4) earthy, and (5) heavy woody-earthy-camphor type of taste and odor. In water pretreatment processes, the addition of chlorine will destroy the taste and odor of the first two types, but will enhance and magnify numbers 3, 4, and 5. If chlorine dioxide or sodium chlorite is added to water containing any of the five stages, either of these chemicals will destroy all five stages. The tastes and odors imparted to waters by both of these groups of organisms have yet to be implicated as disease-causing agents. Thus, they are only aesthetically unacceptable in potable supplies. But as pointed out earlier, a strong earthy odor can be imparted to food and is unpleasant in soft drinks.

Coagulation or flocculation is accomplished by the addition of alum (aluminum sulfate), or ferric chloride to the water. Between the pH's of 5 and 7, these chemicals form a gel-like floc suspended in the water. The chemicals are mixed rapidly in the water and then slowly agitated in the flocculation tank. If softening is required at the same time, lime (calcium oxide) is added with the coagulant. Following floc formation, the water stands in sedimentation tanks for a time to allow settling of the floc. The gel-like consistency of the floc provides a trap for particles, settling these to the bottom of the tank. Adequate treatment by flocculation results in removal of practically all of the suspended material in the water, although it cannot be depended upon for complete removal of disease-producing organisms. Bacteria are carried to the bottom of the tank by the settling floc; however, there is no lethal effect in the action of the floc on the microorganisms. Such organisms will remain alive and may increase in numbers, depending upon the temperature and the types of organisms present as well as the amount of dissolved nutrient material available. As long as these are present and living, it is possible that potentially dangerous cells will be resuspended and carried to the consumer in the water.

The water on top of the settling tank is continuously decanted and put through sand filters which remove the remaining floc. The filter bed may be either a slow or rapid filter. The difference in the two is the degree of coarseness and depth of sand, which allows a different rate of flow through the filter. Filtration further clarifies the water by removal of suspended organic matter, and will also remove microorganism cells. The efficiency of filters varies; however, most often, there will be very little suspended organic matter and extremely few bacterial cells remaining following filtration. An additional purification step insures potability of water.

Sand filters may be augmented in some treatment plants by additional materials such as activated carbon. Some dissolved chemicals will be removed by this added filter, including some types of disagreeable taste- and odor-causing compounds. The purpose for adding activated carbon to a filter in the past has been for aesthetic improvement and not to improve the biological purity of the water. Use of this additional step is determined locally on the basis of the nature of the existing water supply.

The final step in water purification is a chemical treatment of the water with a disinfectant. Some plants provide a chlorine treatment prior to filtration, as mentioned earlier. This addition begins killing microorganisms and helps to prevent growth on the filters. Again, dependent on local conditions, this may be very beneficial. In any case, however, following filtration the clarified water is chlorinated to disinfect and to provide a residual of chlorine that will retard the growth of any microorganisms that enter either the storage facility or the water distribution system.

The requirement for post-treatment chlorination varies in different states. For example, the Texas requirement is to provide that a minimum of 0.2 mg/l free residual chlorine after 20 minutes contact time or greater than 0.20 mg/l combined residual after three hours of contact time has been maintained.

The initial killing of organisms following filtration is important, but it is equally important that some chlorine residual remain in the distribution system. Regardless of the effort and money put into a water distribution system, eventually there will be breaks, leaks, and so on, which will permit the entrance of microorganisms. The presence of a chlorine residual, while not of sufficient concentration to kill these cells, inhibits the growth of the cells and will in this way hold down the numbers that may be present. Because infections are dependent to some extent on the number of cells entering the body, this inhibition of growth is of great benefit, and certainly such retardation of growth would help to limit outbreaks of a number of infections.

The final quality of water that has been put through the coagulation and softening processes varies between treatment plants only in the inorganic chemical content. These treatment processes have little effect on chemicals such as nitrate or sodium. And sodium, for example, is becoming increasingly important from a public health point of view. The habitual intake of sodium on a per capita basis is between 3,000 and 4,000 mg/day in the United States. The sodium in water represents less than ten percent of the total intake. Persons on low sodium diets usually are restricted to less than 500 mg/day. And because it is not completely feasible to reduce food sodium intake below 440 mg/day, the remaining 60 mg/day suggests that some type of demineralized water be consumed.

As far as softening of the water is concerned, the current trend appears to be away from significant reduction of water hardness: around 75 to 100 mg/l is an appropriate hardness value. Apparently, epidemiological studies show a significant increase in cardiovascular disease in the population in areas having very soft water.

The finished water in the distribution system may be potable but not meet the requirements of various food industries. Table 6.2 contains data that compare the desired maximum levels of certain water constituents for two food industries. Most often, industries provide further treatment internally after water is received from the municipal supply.

Chlorination of finished water has, since 1978, been proven to contribute substantially to the formation of chlorinated hydrocarbons.

Table 6.2.   Some Quality Characteristics of Water Used by Two Food Industries

Characteristic	Food Industries	
	Food Canning	Soft Drink
Alkalinity	300	85
Hardness	310	(a)
Calcium	120	—
Chloride	300	500
Sulfate	250	500
Iron	0.4	0.3
Manganese	0.2	0.05
Silica	50	—
Nitrate (NO_3)	45	(a)
Organics, CCE	0.3	0.2
Color	5	10
Dissolved solids	550	(a)

Source: *Water Quality Criteria*, 1968.

Units are mg/l; values that should not be exceeded. (a) controlled internally by treatment.

Some of these compounds have been identified as being carcinogenic in laboratory animals. The trihalomethanes have been specifically identified as being part of the overall group of chemicals produced. Trihalomethanes (THM) include chloroform, dibromochloromethane, and bromoform. The safe exposure level, or MCL, has been recently established at 100 ppb. To combat their occurrence in public supplies, it has been proposed that communities over 75,000 population install activated carbon granular filters, in conjunction with sand filters. This should reduce the THM to safe levels. Replacement and rejuvenation of the carbon is expensive and after water leaves the treatment plant, and is in the distribution system, many other compounds can be formed if chlorination is practiced. Carbon filters do not, therefore, completely solve all problems.

TESTING FOR SANITARY QUALITY OF WATER

Water treated by standard methods should be potable, or safe for consumption from a health standpoint. Obviously, however, there can always be failures of mechanical equipment, as well as human error and neglect in any operation. To guard against these possibilities, federal and state laws require routine testing of water supplies to regularly establish the efficiency of treatment.

Microorganisms that can live, or establish an infection in the gastrointestinal tract of humans or other animals may be likely to get into all surface water supplies. Because of the wide variety of bacteria, fungi, and protozoans that may be included in this category, it would indeed be a difficult task to test for many species individually, particularly because the pathogenic organisms are most likely to be in the minority as compared to the total number present. For example, it would be most improbable that every reasonably sized water sample (100 ml) in a water source would contain sufficient numbers of *Salmonella typhosa* that this organism could regularly be isolated from a contaminated supply. The difficulties of isolation, cultivation, and identification of pathogenic organisms make it impractical to test for them individually. Therefore, a means to indicate the probability of presence of pathogenic organisms in water has been established after thorough testing. This has been accomplished by selection of an indicator organism, the presence of which is assumed to mean that the water has been subjected to fecal contamination. According to the National Research Council, to be adequate indicators of fecal pollution, the organism must be one that is

(1)   applicable to all types of water;
(2)   present in sewage and polluted water when pathogens are present;
(3)   present in numbers correlated with the amount of pollution;
(4)   present in greater numbers than are pathogens;
(5)   not capable of aftergrowth in water;
(6)   capable of greater survival times than pathogens;
(7)   absent from unpolluted water;
(8)   easily detected by simple laboratory tests;
(9)   in possession of constant characteristics; and
(10) harmless to man and animals.

The species selected as the best indicator for use in the United States is *Escherichia coli*. Some strain of *E. coli* is a predominant bacterial species in the intestinal tract of most humans. Its presence is easily indicated by rapid, simple culture techniques, and its identification is readily established by further simple cultural procedures. Testing procedures may vary to some degree in different laboratories because there are some approved alternate standard methods. Approved methods, including complete descriptions of required equipment and culture media may be found in *Standard Methods for the Examination of Water and Wastewater*, 16th edition (1985).

The usual procedures test for organisms closely related to *E. coli*, or the general group termed *coliforms*. Standard tests begin with the Presumptive Test. In this procedure, a series of water samples are cultured in a selective broth that will support the growth of coliforms, but that is inhibitory to many other organisms normally found in water. Growth of bacteria with the production of acid and gas within forty-eight hours is presumptive evidence of the presence of coliforms, and further testing is indicated.

The Confirmed Test is performed by transfer of bacteria, from tubes of broth used in the Presumptive Test that are positive for acid and gas production, to additional selective broth culture media and to differential semisolid agar culture media. Again, if there is formation of gas in the broth cultures and if typical coliform colonies are present on the agar differential media, the test is considered to be positive.

The Completed Test is used following a positive Confirmed Test. This is accomplished by transfer of bacteria from positive confirmed cultures to additional differential agar culture media, and selective broth media and observing for typical coliform colonies and the production of gas. From these cultures, stains are made to determine that the organisms present are of the typical staining type and have morphological characteristics for coliform bacteria. If so, the Completed Test is considered positive, the water is reported to contain coliform bacteria, and therefore is considered to have been polluted by fecal contamination.

Besides the preceding technique, the membrane filter plate count technique is an acceptable alternative, and is more rapid. Additional testing may be done to determine more accurately whether the coliforms are from animal fecal contamination, or possibly from some other, less dangerous source. These tests use special selective culture media, an elevated temperature to distinguish sources of coliforms, and are most useful as a more specific diagnostic measure.

Coliform concentrations allowable in drinking water have in past years been at or below 2.2 per 100 ml, but regulations referred to earlier have established, for the membrane filter test, the following not-to-be-exceeded levels: (a) 1 per 100 ml, arithmetic mean of all samples in a month; or (b) 4 per 100 ml in more than one sample when less than twenty samples are examined per month; or four per 100 ml in more than five percent of the samples when twenty or more are examined per month.

CHEMICAL ADDITIVES IN WATER

Local municipalities and water districts have, in some cases, elected to add fluoride to treated water. This addition is not made for the purpose of purification and the concentration added is not sufficient for killing living cells. Instead, it provides a benefit to the consumer by improving the condition of the teeth. In numerous studies, fluoride has been shown to improve dental health, particularly in the young, and the addition of this chemical to water supplies where it is not found naturally is an inexpensive, convenient method to broadly apply this preventive medicine without any evidence of detrimental effects.

In some areas, water users may complain about red tap water. This color is imparted to the water by iron coming from pipes in either the home or the distribution system. Red water can be controlled easily by the addition of sodium hexametaphosphate (glassy phosphate). At concentrations of about 1 mg/l, this chemical coats the pipe interior and prevents the dissolution or suspension of iron.

THE FUTURE OF WATER SUPPLIES

Although the earth possesses a bountiful supply of water, it is likely that many densely populated areas of the world will face a water supply crisis within a few years. Humans encounter difficulty in the location of a water supply at the point of need, and in keeping that supply unpolluted. Surface water location is generally dependent upon local rainfall, and a challenge for the future is to provide the economically feasible capability to move water from an area of overabundance to an area of need. Periodic droughts emphasize to the human that water sources are not always easily arranged at the particular location desired. In some localities, transfer to the desired location is routinely being made from water sources located on the surface several hundred miles away. With increases in needs for agriculture, industry, and the ever-enlarging human population, however, many areas face a crisis in water supply and location in future years.

Subsurface water, or groundwater, has been used in semiarid locations for general consumption and for agricultural irrigation purposes. In some areas, this use has far outstripped replacement of the groundwater by natural processes. Continued productivity and efficient use of these lands will, in the future, require the development of methods for more rapid movement of water back into these underground reserves. Storage or placement of water in these underground locations is in many ways preferable to storage on the surface. Underground storage reduces loss from evaporation, and helps in large measure to prevent pollution of the supply.

After use, water becomes a problem as a waste and generally is disposed of as sewage. In most localities, this means that the waste water is dumped into rivers, streams, and/or lakes, and often simply flows away to a different location, again producing a problem in location of the water supply. Eventually this wastewater will reach the ocean, evaporate, and be recycled as rainfall to become ground and surface water in some locality. A major challenge for the future is the development of methods of treatment of wastewater that will make it reuseable locally without moving it to a different location hundreds or thousands of miles away, and without being dependent on the uncertain natural recycling process. As will be discussed in more detail in the following chapter, the development of this capability will not only aid greatly in water supply, but it will add immeasurably to prevention of pollution and efficient disposal of wastes.

BIBLIOGRAPHY

AMERICAN PUBLIC HEALTH ASSOCIATION. Benenson, A. S., ed. 1985. *Control of communicable diseases in man.* 14th ed. New York: Author.

AMERICAN PUBLIC HEALTH ASSOCIATION. 1985. *Standard methods for the examination of water and wastewater.* 16th ed. New York: APHA in conjunction with American Water Works Association and Water Pollution Control Federation.

ANONYMOUS. 1976. National interim primary drinking water standards. *Journal of American Water Works Association.* 68(2):57–68.

CENTER FOR DISEASE CONTROL. 1975. *Foodborne and waterborne disease outbreaks, annual summary.* (Issued September 1976). U.S. Department of Health, Education and Welfare. USPHS Publication no. (CDC) 76-8185.

CENTER FOR DISEASE CONTROL. 1985. *Surveillance. Water-related disease outbreaks.* U.S. Department of Health, Education and Welfare. DHHS Publication no. (CDC) 99-2510.

CULP, G., ed. 1984. *Trihalomethane Reduction in Drinking Water* (Pollution Technology Review No. 114). Park Ridge, NJ: Noyes Publications.

INTERNATIONAL ASSOCIATION OF MILK, FOOD, AND ENVIRONMENTAL SANITARIANS, INC. 1979. *Procedures to investigate waterborne illness.* Ames, IA.

NATIONAL ACADEMY OF SCIENCES. 1977. *Drinking water and health,* vol. 1. Safe Drinking Water Committee. National Research Council. Washington, DC.

NATIONAL ACADEMY OF SCIENCES. 1980. *Drinking water and health,* vol. 2. Safe Drinking Water Committee. National Research Council. Washington, DC.

NATIONAL ACADEMY OF SCIENCES. 1980. *Drinking water and health,* vol. 3. Safe Drinking Water Committee. National Research Council. Washington, DC.

NATIONAL ACADEMY OF SCIENCES. 1982. *Drinking water and health,* vol. 4. Safe Drinking Water Committee. National Research Council. Washington, DC.

NATIONAL ACADEMY OF SCIENCES. 1983. *Drinking water and health,* vol. 5. Safe Drinking Water Committee. National Research Council. Washington, DC.

RAND, G. M., and PETROCELLI, S. R., eds. 1985. *Fundamentals of aquatic toxicology.* New York: Hemisphere Publishing Corp.

TARTAKOW, I. J., and VORPERIAN, J. H. 1981. *Foodborne and waterborne diseases.* Westport, CT: AVI Publishing Company, Inc.

7

Wastes and Sanitation

Disposing of wastes without polluting the environment is one of the most difficult tasks of the food industries. The problem is complicated by the broad chemical and physical spectrum of wastes from different types of foods. In a food processing or production plant, the adequate disposal of wastes must be a first consideration; and certainly in a food-service establishment, aesthetics and sanitation demand adequate and proper disposal of waste materials. Proper waste disposal in the food industry may be costly, and detailed planning should go into this aspect of facility location, construction, and operation.

Pollution of any environment is, quite simply, the accumulation of some unused material in a location where it is not found naturally. In and of itself, a pollutant is not necessarily harmful, although some pollutants are toxic or harmful by-products of industry, and others are potentially dangerous living organisms. When not accumulated in large quantities, these are generally termed *contaminants,* and are of primary concern to sanitary science in all locations. The accumulation of pollutants in any unnatural location may be assumed to lead to potential contamination of that location's environment, at least indirectly, by altering both its chemical and the biological makeup.

Wastes that create pollution problems may be characterized as industrial, residential, or agricultural. Each type may be further subdivided into gaseous, liquid, or solid wastes. The source of wastes and the type determine the special problems encountered in disposal without adding to the probability of either contamination or pollution. Food industries are concerned with all of the sources and types of wastes either in the production, processing, serving, or consumption states of operation of food establishments. This broad spectrum of concern presents food industries with perhaps greater problems than some other industries must face.

INDUSTRIAL WASTES

For discussion of pollution and sanitary effects of waste disposal, industrial establishments may be divided into food and non-food industries. Non-food industry wastes may be made up of any type of materials, and all such wastes are known to pollute the environment without proper measures for disposal. Concentration of industries in localized areas creates many of the disposal problems. Most industries do not create gaseous wastes, which are directly toxic in small amounts, but the additive effect of many industries disposing various gases creates toxicity over a period of time, particularly when combined with certain weather conditions. Food industries must be concerned with these wastes from the standpoint of personnel health and well-being, as well as from the standpoints of food processing, in which chemicals may be absorbed by foods to change tastes, odors, or appearance and in food production, when animal or plant crops may be affected by the gaseous pollution.

Solid waste from non-food industries may be of only indirect concern to food industries in that these wastes are likely to be accumulated locally at some place away from the food establishment. Some disposal practices, such as open trash dumps or dumping solid wastes into lakes, harbors, and so on may indirectly create problems for food industries. In addition to aesthetic damage to the environment, the disposal of trash in open dumps creates breeding grounds for flies, other insects, and rodents that may then contaminate a food source. Disposal into bodies of water adds to the pollution of water supplies and alters the aquatic environment in such a way that it interferes with production of aquatic foods.

Liquid sewage from industrial establishments may contain few damaging biological pollutants, but is more likely to contain chemical pollutants that will damage the aquatic environment into which it is disposed. From the standpoint of the food industry, this damage may result in lowered aquatic food production, undesirable tastes and odors in water used in food processing or preparation, or toxicity, either direct or cumulative. These considerations point up the concern food industries must have with water pollution regardless of source or type.

The food industries must be involved, in waste disposal, with prevention of pollution of any environment as well as with the disposal of waste regularly and efficiently to prevent contamination of food products. In actuality, food industry wastes must be considered as including agricultural wastes resulting from food production, whether of plant or animal nature. By far the majority of agricultural waste, by weight, is animal rather than plant in nature. Solid wastes resulting from animal production is a problem of immense magnitude. The location of animal production facilities in rural areas requires, in many cases, that the producer construct, maintain, and operate solid waste treatment and disposal facilities. These facilities must (a) avoid polluting the air in the form of odors, (b) ensure that the animal waste not be dangerous from the standpoint of spread of disease to other animals or humans, and (c) provide some means for recycling, use, or disposal of the treated waste in order that accumulated masses do not in themselves become problems. Of these requirements, the first two are vital to sanitation. Odors will attract insects, inviting potential contamination of food products in a broad area.

Perhaps the magnitude of the food animal production problem can be better appreciated if the following data are considered. Steele and Pier reported that in 1972, the U.S. production, in millions of animals was cattle—110; sheep—20; hogs—110; and poultry—3,000. The total animal solid waste production per year was estimated at 1.3 billion tons. These data, of course, do not include the wastes produced in processing operations. The meat packing industry alone accounts for over 825 million tons of waste per year. That amount is more than twice the tonnage of urban refuse produced per year in the United States. With that type of processing waste production, it can be easily seen that sanitation in the food industry is of utmost concern.

Smith, in 1978, characterized the complex problems of food industry waste disposal by utilizing examples from the potato processing industry. He reports that 1,200 gallons of water are required for each ton of raw potatoes processed, and that this results in 20,000 mg/l of suspended solids in the processing water. Processing in this case then causes the release of a waste effluent with a pH of 6.8, a chemical oxygen demand (COD) of 400 (mg/l), and a biochemical oxygen demand (BOD) of 220 (mg/l). (Chemical oxygen demand is a measure of the oxygen required to remove the chemical load from the water. Biochemical oxygen demand is the oxygen required to remove the biological load from the water.) Such an effluent will cause considerable pollution problems unless it is treated before release to surface waters.

All food processing plants have rather specialized solid and liquid waste disposal problems because in addition to the usual trash and clutter of containers, and so on, unusable portions of plant or animal foods, and cleanup water, contain such high concentrations of nutrients that spoilage by microorganisms, attraction of insects and rodents, and possibly even growth of pathogenic organisms are constant potential dangers. Liquid wastes containing high nutrient content, when disposed in sewage emptying into water supplies, result in pollution by enrichment of microbial populations present in the water. Such enrichment upsets the natural balance and the resulting biologic conditions in the water may be totally undesirable. Examples of various liquid waste nutrient concentrations and respective volumes produced per unit of product are presented in Table 7.1. Ideally, nutrient content of solid and liquid food wastes should be recycled to further enhance useful food production. Such possibilities are being studied in many projects. The magnitude of the waste produced in food processing plants makes it imperative that solutions be sought and that practices be perfected to efficiently dispose of these wastes without danger to the industry, the public, or the environment.

Overall, the food industry is one of the largest water users in any country. Billions of gallons of water are used annually for all aspects of processing, cooling, and other uses. Table 7.2 shows the number of

Table 7.1. Volume, BOD, and Suspended Solids of Some Food Processing Wastes

Commodity	Volume (in Gal. per case)	5-Day BOD[1] (ppm)	Suspended solids (ppm)
Apples		1,700–5,500	300–600
Apricots	57–80	200–1,000	260
Asparagus	70	20–100	30–180
Beans, baked	35	900–1,400	220
Beans, green or wax	26–44	160–600	60–150
Beans, kidney	18–22	1,030–2,500	140
Beans, limas, dried	17–22	1,740–2,880	160–600
Beans, limas, fresh	50–257	190–250	420
Beets	27–70	1,580–7,600	740–2,220
Carrots	23	520–3,030	1,830
Citrus	1,000[2]	1,000–5,000	1,200
Corn, cream style	24–29	620–2,900	300–670
Corn, whole-kernel	25–70	1,120–6,300	300–4,000
Cherries, sour	12–40	700–2,100	20–600
Cranberries	4	2,250	100
Grapefruit	5–56	310–2,000	170–280
Meat-packing house	2,000–8,000[3]	600–1,600	400–720
Milk-processing industry	3–5[2]	20–650	30–363
Mushrooms	6,600[2]	80–850	50–240
Peaches	1,300–2,600[2]	1,350–2,240	600
Peas	14–75	380–4,700	270–400
Peppers		600–1,220	
Potato chips	4,000[2]	730–1,800	800–2,000
Potatoes, sweet	82	1,500–5,600	400–2,500
Potatoes, white		200–2,900	990–1,180
Poultry-packing industry	1½[4]	725–1,148	769–1,752
Pumpkin	20–50	1,500–11,000	785–1,960
Sauerkraut	3–18	1,400–6,300	60–630
Spinach	160	280–730	90–580
Squash	20	4,000–11,000	3,000
Tomatoes	3–100	180–4,000	140–2,000

Source: Potter (1978).

[1]Biological Oxygen Demand.
[2]Per Ton.
[3]Per Gallon of Milk.
[4]Per Chicken.

Table 7.2. Water Use by Various Food and Related Products Industries In 1963

Industry	Total	Quantity of Intake Water, in 10^6 gal.				
		Under 1*	1–9*	10–19*	20–99*	100 and over*
Meat products	2,734	1,319	695	186	343	191
Dairies	4,638	1,963	1,492	452	577	154
Canned and frozen foods	2,340	1,161	502	171	335	171
Grain mills	2,158	1,689	313	61	60	35
Bakery products	2,608	1,685	653	124	129	17
Sugar	174	32	11	6	19	106
Candy and related products	638	407	146	29	45	11
Beverages	3,418	1,522	1,241	240	278	137
Miscellaneous food and related products	2,890	1,597	701	217	230	145

Source: Handbook of Environmental Control. 1973.

*Example: 1,319 industries used $<1 \times 10^6$ gal./year

industries that use volumes in the millions of gallons of water annually. In considering these data, it must be remembered that not all of the water is turned into wastewater. Many industries use a once-through cooling process, while others have the capability to recycle the water internally for repeated use.

Foodservice wastes are essentially like residential wastes; however, most ordinances, including the recommendations in the *Food Service Manual* (1976) are quite specific as to how this garbage is to be handled, stored, and disposed. The recommendation of the above reference reads:

GARBAGE AND REFUSE

6-601   CONTAINERS.

(a) Garbage and refuse shall be kept in durable, easily cleanable, insect-proof containers that do not leak and do not absorb liquids. Plastic bags and wet-strength paper bags may be used to line these containers, and they may be used for storage inside the food service establishment.

(b) Containers used in food preparation and utensil washing areas shall be kept covered after they are filled.

(c) Containers stored outside the establishment, and dumpsters, compactors and compactor systems shall be easily cleanable, shall be provided with tight-fitting lids, doors or covers, and shall be kept covered when not in actual use. In containers designed with drains, drain plugs shall be in place at all times, except during cleaning.

(d) There shall be a sufficient number of containers to hold all the garbage and refuse that accumulates.

(e) Soiled containers shall be cleaned at a frequency to prevent insect and rodent attraction. Each container shall be thoroughly cleaned on the inside and outside in a way that does not contaminate food, equipment, utensils, or food preparation areas. Suitable facilities, including hot water and detergent or steam, shall be provided and used for washing containers. Liquid waste from compacting or cleaning operations shall be disposed of as sewage.

6-602   STORAGE.

(a) Garbage and refuse on the premises shall be stored in a manner to make them inaccessible to insects and rodents. Outside storage of unprotected plastic bags or wet-strength paper bags or baled units containing garbage or refuse is prohibited. Cardboard or other packaging material not containing garbage or food wastes need not be stored in covered containers.

(b) Garbage or refuse storage rooms, if used, shall be constructed of easily cleanable, nonabsorbent, washable materials, shall be kept clean, shall be insect-proof and rodent-proof and shall be large enough to store the garbage and refuse containers that accumulate.

(c) Outside storage areas or enclosures shall be large enough to store the garbage and refuse containers that accumulate and shall be kept clean. Garbage and refuse containers, dumpsters and compactor systems located outside shall be stored on or above a smooth surface of nonabsorbent material such as concrete or machine-laid asphalt that is kept clean and maintained in good repair.

6-603 DISPOSAL.

(a) Garbage and refuse shall be disposed of often enough to prevent the development of odor and the attraction of insects and rodents.

(b) Where garbage or refuse is burned on the premises, it shall be done by controlled incineration that prevents the escape of particulate matter in accordance with law. Areas around incineration facilities shall be clean and orderly.

The Food Ordinance of the City of Houston, Texas differs very little from these recommendations. The reasoning behind the recommendations and ordinances is that garbage must be kept and disposed of in such a way that the areas are easily sanitized, and that insects, rodents, and other pests are not attracted.

RESIDENTIAL WASTES

Residential wastes are usually either liquid sewage or solid garbage, including paper, plastic, glass, or metal containers. Sewage is generally treated in septic tanks for individual residences or in sewage treatment disposal plants for municipalities. Such sewage contains human excrement; chemicals such as detergents, pesticides, and oils; and animal and vegetable wastes from foods. This mixture combines the potentialities of infectious disease organisms, stable chemicals that may be concentrated in plant or animal tissues, chemicals that may be directly toxic to plants or animals, and nutrients that will selectively enrich and/or inhibit growth of microorganisms that could bring about degradation of the solids present. These possibilities require that residential sewage be treated to reduce potential pollution and contamination possibilities. Treatment generally will include some combination of (a) sedimentation of solids, (b) digestion of solids and removal of some nutrients by microorganisms, and (c) in some cases, chemical treatment with chlorine to kill potential pathogens. Following treatment with some combination of the preceding, municipal sewage is generally emptied into streams, rivers, lakes, reservoirs, or possibly directly into the ocean. Without adequate treatment, it is obvious how rapidly pollution of the water will occur, including contamination of the water with disease-producing organisms. If treatment is truly adequate, then digestion of remaining solids and organic materials should be completed in the receiving water and the natural biologic and chemical balance of the water should not be greatly altered. This is the ideal that must be attained to preserve our water supplies and environment. To reach this satisfactory state, all segments of the sanitation, water, and ecological sciences must continue to search for and explore potential technologies for application to these problems.

Solid residential wastes present somewhat the same problems as do industrial wastes, with the addition of the variety of solids present. Those solid materials that are easily degradable or burnable do not present as many potential problems as do such materials as plastic and glass. Convenience packaging has resulted in tremendous tonnages of very stable materials which must be removed from the environment in some manner. Urban solid waste management, in most cities, contributes one of the highest budget items of the municipality, and in some cases, it indeed is the highest cost item in the entire budget.

WASTE DISPOSAL

The variety of waste products resulting from human activities has increased with improved standards of living, increased population, and increased concentrations of population. These wastes must be disposed of in such a manner as to (a) present no danger to human health or well-being, (b) present no damage to plant or animal productivity, and (c) present no damage to the natural function of all parts of the environment, including its natural beauty. If these objectives could be attained, waste disposal would no longer be the pressing concern it now is. However, the cyclic processes of nature might still be in imbalance; and in the long course of events, this imbalance could create some deficiency in the productivity and function of life. Waste disposal must be considered in terms of degradation of the waste to usable raw materials needed for recycling. These end results will permit maintenance of the balance of nature, and an equilibration of the environment that will permit most efficient and profitable life for the human.

The volatility of gaseous materials perhaps makes this type of waste most difficult to dispose. The prevention of air pollution has been approached from two directions: (a) alter procedures to avoid production of the air pollutant, and (b) prevent release of the air pollutant by adsorption onto a solid or absorption into a liquid, or by conversion of the pollutant into a solid or liquid. Changing processes or fuels can control the type of gas or chemical air pollutant produced. An example of such control is observed in reduction or removal of lead from gasolines. Adsorption of chemicals or smokes onto filters prevents or reduces the escape of these materials into the air, and the filter can then be handled as a solid waste. Many industries have already instituted processes for reducing air pollution; but improved methods must constantly be developed.

Liquid waste disposal must be considered both from the standpoint of removal and degradation of dissolved chemicals and of reclamation of the water. Many soluble wastes in the past have been reclaimed as useful by-products in some industries. This remains a possibility in many industries today, if processes can be altered to change some of the by-products that are unusable to materials of value. The study of liquid wastes to determine degradability, preferably by natural biological means to avoid production of additional waste products, must be a major preoccupation of persons concerned with waste disposal. To do this, much additional knowledge of natural basic biological processes is required. One experimental method for disposal of liquid wastes, including that containing suspended solids, that has been used is a system of spraying the waste over soil surfaces. This is done in the expectation that soil organisms will degrade the dissolved materials, converting them to usable raw materials which then become available to other microorganisms and plants. If a balanced system can be detected and the recycling balance maintained, this process holds much promise for the disposal of many industrial and agricultural waste products.

A short drive for most Americans in any direction brings one face to face with a huge solid waste problem in the form of old wrecked automobiles. This problem seems only to grow, never to recede, in spite of the fact that one obvious solution in this case is reclamation of much of the metal in this trash. With the rate at several million cars disposed of per year, the time is approaching when it will be essential that we begin to solve the problem of this unsightly blot which only provides shelter for animals and pests and generally mars the environment.

Industrial solid wastes, as mentioned earlier, are a problem compounded by concentration of industries. Well over 100 million tons of industrial solid wastes are produced each year excluding the agricultural industries, which add over 2 billion tons of wastes each year. The magnitude of the human public health and animal health problems posed by this tonnage of potential contamination (if improperly handled and disposed of) is almost beyond prediction. To compound the problems in this case, with growing urban needs for food, an increasing proportion of this waste is now generated in or near urban areas.

The almost casual attention given waste disposal in the past has given way to urgent attempts at planning and action in the present. In addition to the ecological considerations and the preservation of the beauty of environments, the public health and sanitation considerations involved in the problems of waste disposal make it mandatory that individuals, industries, and all levels of government seek more efficient, economically feasible, and effective methods of waste disposal. The urgency of the situation has given rise to a new profession—waste management. In the case of solid wastes, at least, a systems concept has evolved that involves consideration of social, economic, public health, engineering, law, urban planning, rural planning, and geography factors; and disposal of waste materials to the best interest of all, rather than the most convenient disposal for the producer of the waste. The future growth, expansion, and productivity of all industries, including food industries, depends upon seeking, finding, and putting into operation the most effective methods for management, treatment, and disposal of all types of wastes from all sources. No longer can any individual, company, or governmental unit wait for a solution to be forced upon them. Waste management in many forms has entered the picture as a new sanitary science that becomes more and more closely allied to sanitation in food industries as well as to other industries.

BIBLIOGRAPHY

AMERICAN PUBLIC HEALTH ASSOCIATION. 1985. *Standard methods for the examination of water and wastewater,* 16th ed. APHA, in conjunction with American Water Works Association, and Water Pollution Control Federation. New York: Author.

AMERICAN PUBLIC HEALTH ASSOCIATION. 1985. Benenson, A. S., ed. *Control of communicable diseases in man.* 14th ed. New York: Author.

BOND, R. G., STRAUB, C. P., and PROBER, R., eds. 1973. *Handbook of environmental control.* vol. III. Boca Raton, FL: CRC Press, Inc.

CITY OF HOUSTON, CITY COUNCIL. 1978. *The Food Ordinance.* Houston, TX: City of Houston Health Department.

GUTHRIE, R. K., and DAVIS, E. M. 1985. Biodegradation in effluents. In *Advances in biotechnological processes,* eds. A. Mizrahi and A. L. van Wezel. New York: Alan R. Liss, Inc.

JOHNSON, H. D., ed. 1970. *No deposit—No return.* Anthology of papers presented at the 13th National Conference of the U.S. Commission for UNESCO. Reading, MA: Addison-Wesley Publishing Co.

POTTER, N. N. 1986. *Food science.* 3rd and 4th eds. Westport, CT: AVI Publishing Co., Inc.

SMITH, O. 1978. Waste treatment. In *Encyclopedia of food science,* eds. M. S. Peterson and A. H. Johnson. Westport, CT: AVI Publishing Co., Inc.

STEELE, J. H., and PEIR, S. M. 1972. *Conversion of animal wastes to fuel.* Proceedings of U.S. Animal Health Association Meeting. Richmond, VA.

U.S. CONGRESS. 1970. *Environmental quality.* Washington, DC: U.S. Government Printing Office.

U.S. DEPARTMENT OF HEALTH, EDUCATION AND WELFARE. 1976. *Food service sanitation manual.* Public Health Service. Food and Drug Administration. DHEW Publication no. (FDA) 78-2081. Washington, DC.

8

Regulations Controlling Sanitation in Food Production and Processing

Protection of the public from foods that may be harmful has been a concern of societies since ancient times. One of the earliest laws in England, enacted in 1266, prohibited sale of food that was "not wholesome for man's body." Many cultural and religious practices dealt with assuring wholesome food for societies in ancient times. While there may have been some basis for those practices at the time, current technology permits processing, protecting, and preserving foods in manners that eliminate the dangerous aspects formerly present.

The earliest American laws were enacted by colonists shortly after migration of Europeans to this country began. These laws frequently dealt with the purity of bread. However, many of them were merely fragments of rational laws, and these continued until the first federal law was enacted by the United States in 1848 with the passage of the Import Drug Act. The food supply was not really protected until the Federal Food and Drug Act of 1906. That act actually prohibited interstate commerce in adulterated or misbranded foods, and specified few other prohibitions.

The U.S. Food, Drug, and Cosmetic Act, enacted in 1938, and since amended many times, forms the basis for our current protection of the public against adulterated or dangerous food stuffs. That law includes

A food shall be deemed to be adulterated—(a)(1) If it bears or contains any poisonous or deleterious substance which may render it injurious to health; ... (3) if it consists in whole or in part of any filthy, putrid, or decomposed substance, or if it is otherwise unfit for food; or (4) if it has been prepared, packed, or held under insanitary conditions whereby it may have become contaminated with filth, or whereby it may have been rendered injurious to health; or (5) if it is, in whole or in part, the product of a diseased animal or of an animal which has died otherwise than by slaughter; . . . .

The broad general language used in the federal law allows some flexibility in regulations established by other governmental bodies, yet provides protection to all persons regardless of the action of local governments.

The federal government has jurisdiction over food industries involved in any manner in shipment of foods from one state to another, and federal laws therefore apply to those specific establishments. Because of this jurisdiction, federal law applies in greater degree to food production and processing than to foodservice industries. Therefore, a major portion of regulations dealing with food quality is included in Good Manufacturing Practices (GMPs), which make up Part 128 of the Code of Federal Regulations, published in the *Federal Register*. This document is entitled "Human Foods; Current Good Manufacturing Practice (Sanitation) in Manufacture, Processing, Packing or Holding." The original GMPs were issued in 1969 and revisions have been considered to include specific product GMPs for bottled drinking water, low-acid canned foods, acidified foods, smoked fish, and irradiated foods. GMPs included deal with general definitions, and these are used to determine whether the "facilities, methods, practices, and controls used in the manufacture, processing, packing, or holding of food are in conformance with or are operated or administered in conformity with good manufacturing practices to assure that food for human consumption is safe and has been prepared, packed, and held under sanitary conditions." Specifications for plants and grounds, equipment and utensils, facilities and controls, operations, processes, personnel, exclusions, and natural or unavoidable defects in food are included. Separate sections deal with smoked and smoke-flavored fish and with frozen raw breaded shrimp.

With the increase in recent years in the number of food industry establishments that carry on interstate commerce in foods, the federal government would find it more costly and difficult to adequately control the quality of foods without assistance from state agencies. Most state laws and some local ordinances are also directed toward the regulation of interstate industries as well as those that operate entirely within the state's boundaries, and also apply to food production and processing as well as to foodservice. For example, the Food Ordinance of the City of Houston, Texas of 1978 defines a food establishment as "all food service establishments and food processing establishments." It continues " 'Food processing establishment' shall mean a commercial establishment in which food is manufactured or packaged for human consumption and all commercial establishments bottling water or other liquids intended for human consumption. The term food processing establishment does not include a food service establishment or commissary operation." The broader application necessitates more specific rules in some cases, or these specific items may be left for county or city government regulation. The State of Georgia, in the Rules of Georgia Department of Agriculture Food Division Regulations, has regulations that are applicable to processing plants in addition to those that apply to other food establishments. These regulations deal with floor construction; the manner of water re-use; the size, installation, and manner of cleaning utensils and equipment; and the specifications for all refrigeration and freezing of perishable foods. County and city ordinances may be directed toward food production and processing, but are more likely to concentrate most heavily on spelling out specific requirements for foodservice establishments, while assisting the enforcement of state and federal laws dealing with production and processing. If food production and processing industries comply with the general and/or specific requirements of the state and federal laws concerning food sources, quality, and handling, there will be no difficulty in compliance with the more specific language of any local ordinance.

State laws governing the sanitary and quality requirements of foods may follow the general broad language of the federal law, or may make much more specific regulations concerning specific foods or situations. Examination of such state regulations will reveal examples of both approaches. The State of Texas Food, Drug, and Cosmetic Act is an example of the first approach, with some few additions as illustrated here (for more detail, see Appendix Part 3):

Sec. 3. The following acts and the causing thereof within the State of Texas are hereby declared unlawful and prohibited:

(a) The manufacture, sale, or delivery, holding, or offering for sale of any food, drug, device, or cosmetic that is adulterated or misbranded; . . . .

Sec. 10. A food shall be deemed to be adulterated:

(a) (1) If it bears or contains any poisonous or deleterious substance which may render it injurious to health; . . . or (3) if it consists in whole or in part of a diseased, contaminated, filthy, putrid, or decomposed substance, or if it is otherwise unfit for food; or (4) if it has been produced, prepared, packed or held under unsanitary conditions whereby it may have become contaminated, or whereby it may have been rendered injurious to health; or (5) if it is the product of a diseased animal or an animal which has died otherwise than by slaughter, or that has been fed upon the uncooked offal from a slaughterhouse; or (6) if its container is composed, in whole or in part, of any poisonous or deleterious substance which may render the contents injurious to health.

(b) (1) If any valuable constituent has been in whole or in part omitted or abstracted therefrom; or (2) if any substance has been substituted wholly or in part therefor; or (3) if damage or inferiority has been concealed in any manner; . . . or (6) if it be fresh meat and it contains any chemical substance containing sulphites, sulphur dioxide, or any chemical preservative which is not approved by the United States Bureau of Animal Industry or the Commissioner of Health.

The city or county ordinance will deal with issues in very specific language for the most part, and may very well refer to state and federal requirements, as in the following excerpts from Dallas City Ordinance No. 15578 of 1977 which amended the Food Establishments and Drugs section of the City Code.

Sec. 17-2.1. FOOD QUALITY.

(a) *Sources.* A food products establishment shall not use, serve, sell, or distribute processed food from a source (whether inside or outside the city) that has not been inspected and approved by a federal, state, or local government health authority.

(b) *Adulterated and misbranded food.* A food products establishment shall not use, serve, sell, or distribute adulterated or misbranded food.

(c) *Milk.* A food products establishment shall not:

(1) use, serve, sell, or distribute fluid milk or a fluid milk product that is not grade "A" pasteurized milk or milk product;

(2) sell or distribute fluid milk or a fluid milk product that is not in the sealed container in which it was packaged at the milk plant;

(3) use (except in instant desserts, whipped products, and cooking), serve, sell, or distribute dry milk or a dry milk product that is reconstituted in the establishment; or

(4) use, sell, or distribute dry milk or a dry milk product that is not made from pasteurized milk.

(d) *Shellfish.*

(1) A food products establishment that processes or packs fresh or frozen shucked shellfish for sale or distribution shall pack the shellfish in non-returnable packages identified with the name and address of the original shell stock processor, shucker-packer, or repacker, the kind and quality of shell stock, and the interstate certification number issued by the state or foreign shellfish control agency.

(2) A food products establishment that sells, distributes, or serves shucked shellfish shall keep them in the certified container in which they were packaged for sale or distribution until preparation for serving or display for sale.

(e) *Eggs.* A food products establishment that uses, prepares, sells, or distributes eggs or egg products shall use, prepare, sell, or distribute only U.S.D.A. graded or inspected:

(A) clean whole eggs with shell intact without cracks;

(B) pasteurized liquid, frozen, or dry eggs;

(C) pasteurized dry egg products; or

(D) hard boiled eggs that, if peeled, have been commercially prepared and packaged.

Sec. 17-2.2   FOOD PROTECTION.

(a) A food products establishment shall not handle food in a way that is likely to result in the adulteration of the food.

(b) A food products establishment shall protect food from contamination, including, but not limited to, dust, insects, rodents, unclean equipment and utensils, flooding, drainage, and overhead leakage and drippage.

(c) A food products establishment shall protect processed food that is unpackaged, from coughs, sneezes, and unnecessary handling.

Sec. 17-2.3.   FOOD MANUFACTURE, PROCESSING, AND PREPARATION.

(a) *Fixed location*. Except as otherwise authorized by this chapter or other law, a food products establishment that manufactures, processes, or prepares food shall conduct these operations only at fixed premises and shall prepare food only in a completely housed area inside a fixed facility. However, a catering service may prepare food in an area that is not housed, if preparation does not involve the mixing of more than one ingredient. This subsection does not prohibit the service, sale, or distribution of food outside a fixed facility.

The City of Houston, Texas, Food Ordinance includes the following:

SEC. 19-21.

All food establishments shall comply with the following items of sanitation.

Item 1:

FOOD SUPPLIES; General. Food shall be in sound condition, free from spoilage, filth, or other contamination and shall be safe for human consumption. Food shall be obtained from sources that comply with all laws relating to food and food labeling. The use of food in hermetically sealed containers that was not prepared in a food processing establishment is prohibited.

Special requirements.

(a) Fluid milk and fluid-milk products used or served shall be pasteurized and shall meet the Grade A quality standards as established by applicable laws. Dry milk and dry-milk products shall be made from pasteurized milk and milk products.

. . . .

(c) Only clean whole eggs, with shell intact and without cracks or checks or pasteurized liquid, frozen, or dry eggs or pasteurized dry egg products shall be used, except that hard boiled, peeled eggs, commercially prepared and packaged, may be used.

(d) All frozen desserts such as ice cream, soft frozen desserts, soft serves, ice milk, slush, non-carbonated fruit flavored frozen snow cones, sherberts, and their related mixes shall meet the standards of quality established for such products by applicable laws and regulations.

(e) All food manufactured, processed or packaged in commercial food processing establishments or commissaries shall be labeled according to all applicable laws.

Item 2:

FOOD PROTECTION; General. At all times, including while being stored, prepared, displayed, served, or transported, food shall be protected from potential contamination by all agents, including dust, insects, rodents, unclean equipment and utensils, unnecessary handling, coughs and sneezes, flooding, draining, and overhead leakage or overhead drippage from condensation. The internal temperature of potentially hazardous foods shall be 45 degrees F. *(7 degrees C.)* or below or 140 degrees F. *(60 degrees C.)* or above at all times, except as otherwise provided in this ordinance.

Emergency Occurrences. In the event of an occurrence, such as a fire, flood, power outage, or similar event, which might result in the contamination of food, or which might prevent potentially hazardous food from being held at required temperatures, the person in charge shall immediately contact the health officer. Upon receiving notice of this occurrence, the health officer shall take whatever action he deems necessary to protect the public health.

COMPLIANCE WITH LAWS GOVERNING FOOD PRODUCTION

In food production industries, the major concern in sanitation is that the food produced contains no material harmful to health, either of a biological or chemical nature, and that the process of production offers the least possible opportunity for contamination with anything harmful to human health. Food production industries vary greatly in the amount and kind of inspection and supervision that may be expected from government agents or industry representatives seeking a source of supply. The farmer producing plant foods only may not be inspected by governmental agencies at all, while the large-scale cattle, hog, or poultry producer may expect regular or irregular inspection depending upon the nature of state laws and the type of operation. Those who do commercial fishing may expect regular inspection of their catches depending upon the local conditions and the type of seafood with which they deal. More details on specific requirements are given in chapter 13.

Producers of milk and milk products are also subject to frequent inspection of production and processing facilities, and to regular laboratory examination of products. These food producers are usually governed by separate statues that very specifically detail conditions and processes of operation. The special requirements are discussed in more detail in chapter 9.

In all food processing operations, the requirements and procedures for compliance, together with many of the penalties for non-compliance are spelled out in city and state codes. If a food operation is not subject to a requirement, it will be specifically excluded.

From the Dallas City Code:

"Sec. 17-1.3. GENERAL AUTHORITY AND DUTY OF THE DIRECTOR OF ENVIRONMENTAL HEALTH AND CONSERVATION.

The director of environmental health and conservation or an officer or employee designated by the director, may enforce any city ordinance applicable to a food products establishment. The director may also enforce a state or federal statute or regulation applicable to a food products establishment operating within the city if that enforcement is not contrary to law. The director shall implement and enforce this chapter.

Sec. 17-1.6.   EXCLUDED ESTABLISHMENTS.

This chapter does not apply to:

(1) milk producers, haulers, or distributors operating under a permit issued under Chapter 26 of this Code;

(2) food operations that are regulated and licensed under federal or state law (as illustrated by, but not limited to, day care facilities, nursing homes, and meat processors);

(3) sale, distribution, transportation, or storage of a raw agricultural commodity (including, but not limited to, raw vegetables and fruit, and pure honey) by the original producer;

(4) sale, distribution, or service of food at an event, party, or other special gathering that is not open to persons other than the members or invited guests of the sponsor; or

(5) retail sale or distribution of nonpotentially hazardous food from a fixed facility if the food is acquired and sold or distributed in cans, bottles, or other prepackaged containers that are not opened before obtained by a consumer, and no food manufacturing, processing, or preparing operations are conducted at the facility.

Sec. 17-9.1.   AUTHORITY TO ENFORCE.

The director of the department of environmental health and conservation shall administer and enforce this chapter.

Sec. 17-9.2.   PERMITS.

(a) Requisite. A person shall not operate a food products establishment from a facility (whether fixed or movable) located inside the city without a permit issued by the director. A separate permit is required for each:

(1) separate, fixed facility from which an establishment operates; however, if a facility contains establishments under different ownership, a separate permit is required for each establishment that is under a separate ownership;

(2) vehicle used to operate a catering service and each mobile food unit and vending machine service unit; and

(3) food vending machine facility containing one or more vending machines that dispense potentially hazardous food, if it is located at a facility from which no food products establishment (other than the food vending machine operation) is operated.

In general, regardless of the type of food production, strict compliance with the requirements of federal and state laws will mean that the food produced will be of better quality, and will be produced in greater abundance by avoiding loss due to destruction or disease of the food crop. Food shipping and processing industries should inspect producers' facilities and/or crops prior to harvest as a matter of information dealing with quantity and quality of supply. The producer of food products must consistently keep informed concerning recommended procedures in all aspects of production. The source and type of recommendation will vary with the type of food product, and with local conditions. Only those aspects of production that affect the sanitary qualities of foods are considered here.

Production of plant foods of suitable sanitary quality will be affected by a variety of factors, including the major concerns of prevention of plant disease and damage to the food by insects or other animals. The producer can do much to assure quality products by the use of insect- and disease-resistant varieties for planting. Great strides have been made in breeding plant strains that possess considerable resistance to many plant pests. Although this process will aid in producing high quality plant foods, knowledge in this area has not advanced sufficiently to permit use of these plant varieties as a sole means of assurance of good quality products.

To increase chances of avoiding damage by disease or insects, chemical control agents must be used. The food producer needs current, updated information for each growing season as to the most effective fungicides, bactericides, insecticides, and herbicides available for use on his or her crops. Such current information includes the specific nature of the chemical, the proper timing schedule of application to avoid chemical contamination in excess of allowable tolerances, and recommended methods of application that will be safest for both the applier of the chemical and the user of the product. Following the recommended procedures, the plant food producer will be able to maximize quality and quantity of yield and to avoid any damage to the product resulting from chemical contamination.

An additional factor that will influence the product's sanitary quality when it reaches the user is the timing and method of harvest. Variations, depending upon the product, will require differences in these processes to assure the proper degrees of ripening as combined with handling, packing, and shipping characteristics of the food. Over-ripeness, bruising, or improper handling will not only damage the appearance of the product, but will lead to possible spoilage or growth of potential pathogenic organisms.

Some plant food crops require additional attention due to their nature. For example, grains must be harvested at a specific time to avoid moisture and possible mold contamination. Some molds will produce toxic products that have serious effects in the human, and cause additional damage in the appearance and quality of the grains. Nuts must be harvested in a manner to minimize soil contamination and cracking of shells to prevent potential hazards from pathogenic bacteria.

In harvesting of all plant food crops, the producer must take care to handle and store the harvested foods at a temperature that will maintain quality and that will not increase the growth of contaminating spoilage and/or pathogenic organisms. Following recommended procedures concerning varieties, chemical control and proper harvest and handling will ensure food products of good nutritive and sanitary quality. Producers who regularly follow these practices will soon be recognized for the reliability of their products by food processors, who will continue to seek these high quality products in an attempt to reduce the magnitude of problems encountered in food processing.

Milk producers encounter special problems in ensuring continuing good sanitary quality because of the fact that milk is so easily contaminated by microorganisms and serves as an excellent culture medium for them. The first requirement for the dairy farmer is that he or she obtain and maintain a disease-free herd. Many diseases of cattle are transmissible to humans, and most infections in dairy cattle will eventually result in the infectious organisms being shed in the milk. So, in addition to the fact that the infection will reduce milk production, that milk produced will be contaminated with pathogenic organisms. If the dairy operator obtains disease-free cattle and carries on a regular testing program for tuberculosis, brucellosis, and so on, then such epidemic types of diseases are not likely to present problems. Sporadic infections such as mastitis are most likely; and it is essential that the milk producer keep informed of the most recent preventive and diagnostic measures for such diseases. Clean barns, facilities, and equipment that are regularly and adequately disinfected will prevent most such transmissions. Prompt diagnosis and treatment of cases of infection that do occur prevent production loss and infection spread. Antibiotic therapy is effective in most cases in treating mastitis; however, the animal must be removed from production for some time to avoid excessive amounts of antibiotic residual in milk produced after the infection is cleared. Of course, in no case should the animal be in production until the infectious organism is cleared from the milk.

Currently, the economic aspects of milk production necessitate that most such operations be on a fairly large scale, and in most cases mechanization is essential for economic success. Mechanization presents problems in maintaining sanitary milk production. It is imperative that all barns, machinery, and equipment be completely cleaned and disinfected at regular intervals to avoid culturing microorganisms in or on the equipment (milk provides an inoculum for such culturing of microorganisms). A dairy operator should establish a thorough, complete cleaning and disinfecting routine and then follow that routine faithfully. Periodic, frequent sampling along the production line will spot any breakdown or deficiencies in equipment or routine and will allow correction before great loss occurs.

A major factor in maintaining sanitation in milk production is rapid cooling after milking, and adequate refrigeration during storage and shipment to processing facilities. Rapid cooling of milk inhibits multiplication of those organisms present—and bacteria will always be present in raw milk in some numbers—which will ensure that the milk will remain of good quality for a longer period of time. If the milk is allowed to warm to any degree, growth of microorganisms present will occur, thereby increasing the total bacterial count and reducing the quality of the milk. Refrigeration of the product until it reaches the processing plant is essential.

Producers of meat and poultry must begin with the same prime requirements as for milk production; that is, obtaining and maintaining disease-free herds and flocks. If animals are obtained disease-free, then some fairly simple, routine practices will usually assure maintenance of this condition. For cattle, swine, sheep, and poultry, disease testing programs are available in most areas that will be invaluable to the producer in maintaining disease-free facilities. Where cases of disease are discovered, destruction of diseased animals and prompt, thorough disinfection of facilities, where possible, is essential. If antibiotics are used in disease treatment, or in feeds as growth stimulants, scheduling of such treatment in relation to slaughter is important to avoid residual amounts of the antibiotic in the meat that will be above the tolerance allowed.

In many cases, avoidance of episodes of infectious disease may be a matter of care in choice of feed source for the animals. In swine and poultry production, a particularly dangerous practice is the feeding of raw kitchen garbage, which may contain microbial contaminants that are capable of producing fatal epidemics of diseases in the animals and/or are pathogenic for humans. Animals that appear to be sick should never be marketed for reasons in addition to the ethics and illegality involved. Such animals should be studied for diagnosis in order that treatments may be instituted where practicable, and further losses may be avoided.

Animal production facilities for meat or poultry are in many cases impossible to maintain in a completely attractive sanitary condition.

However, in the case of poultry production, indoor facilities with considerable mechanization in maintenance are becoming more common. In all such production facilities, a major job is disposal of solid wastes. With concentration of production in local areas, this animal waste disposal is a major sanitary problem. Adequate management of this problem is essential for successful production of a satisfactory product in these industries.

Production of fish and seafoods is an operation that must depend heavily on local conditions, both in the presence of the aquatic sources, and on the degree of pollution of these sources. If the water, salt or fresh, is free of pollution, the product coming from it should be a suitable food product from a sanitary quality standpoint. Commercial fishers should check with local health officials regularly and frequently on the pollution of local waters, and should abide by the recommendations of those officials as to safety of fishing areas. Some fish and shellfish have been shown repeatedly to harbor pathogenic microorganisms for long periods, and have been shown to accumulate toxic chemicals in large amounts. It is essential, therefore, that aquatic foods be harvested only from waters that are not polluted either biologically or chemically. The decrease in availability of such unpolluted waters is of great concern. Current efforts toward abatement of pollution must succeed to prevent the tremendous loss of this food source.

Giving proper attention to avoiding biological contamination in order to prevent diseases of plants or animals, and to the correct and timely use of chemical agents of control, will result in the production of raw foods of good sanitary quality that will aid in large measure in maintaining sanitation throughout the food processing and foodservice industries.

The general application of the following rules will keep the processor and consumer assured of good quality raw foods:

(1) Know your supplier.
(2) Visit his or her facility and see your products being produced.
(3) See the supplier's inspection records and observe for correction of faults.
(4) Do not accept delivery of shipments with obvious defects such as gross dirt, wrong temperature, etc.

Most foods can remain in good condition with little chance of being contaminated if harvested at the proper time, kept at cool temperatures, and handled with reasonable care. Of these factors, the wrong temperature will spoil the effects of proper harvest and handling, and will allow gross multiplication of any contamination present.

FOOD PROCESSING

The food processor must observe and comply with those sections of federal and state laws, as quoted previously, that deal with food production, and must also maintain compliance with other regulations of these governmental agencies as well as any local ordinances governing practices and conditions in food industries. Those general rules specified earlier will start the processor in good sanitary condition, and will permit avoidance of many potential trouble areas. Some state and local laws require that sources of raw foods be approved by those government agencies before raw foods can be used in processing. The function of food processing, in a sanitary sense, is to take a raw food of good quality and make it cleaner, more desirable, and more wholesome by treating it in such a way that it is made ready for preparation for service and will remain in the best possible condition while it is stored, shipped, or being readied for preparation. When good quality food is received, it is a major task of the processor to ensure that it is in no way contaminated, biologically or chemically, during storage, handling, or shipping. Toward this end, governmental regulations are established as discussed in the following.

The U.S. Food, Drug, and Cosmetic Act is a law written to provide broad coverage for setting standards of quality and identity, cleanness and sanitation of food products. Specific instructions concerning details of practices that can or cannot be approved are not detailed in most instances, but may be in the form of (a) tolerances for certain specific chemicals, (b) permissible microbial counts for some foods, or (c) specific prohibition of the use of certain materials as food additives or preservatives.

Some state regulations are also written with the same broad coverage intent; however, some more specific do's and don'ts are included in state regulations for food processing establishments. Municipalities may well group processing and service establishments into a single category for general regulations, and in separate, more specific rulings single out certain types of food establishments that have specialized requirements. For example, the Food Ordinance of the City of Houston, Texas contains the following definitions:

"Food" shall mean any raw, cooked, or processed edible substance, ice, beverage or ingredient used or intended for use or for sale in whole or in part for human consumption.

"Food establishment." The term "food establishment" shall mean and include all food service establishments and food processing establishments.

"Retail food store." "Retail Food Store" shall mean any establishment where food and food products are offered for sale to the ultimate consumer and intended for off-premise consumption; provided, however, the term shall not include an establishment where food is primarily prepared and sold for individual portion service.

"Potentially hazardous food." The term "potentially hazardous food" shall mean any food that consists in whole or in part of milk or milk products, eggs, meat, poultry, fish, shellfish, edible crustacea, or other ingredients, including synthetic ingredients, in a form capable of supporting rapid and progressive growth of infectious or toxigenic microorganisms. The term does not include clean, whole, uncracked, odor-free shell eggs or foods which have a pH level of 4.5 or below or a water activity (Aw) value of 0.85 or less.

Regardless of the generalities or specifics of various laws, enforcement may well follow a concept of good practice, which will provide for some flexibility within a firm, common-sense, framework of safety from the viewpoint of sanitation. The language of federal and many state laws in the use of terms such as *filth, putrid, decomposed, adulterated, unfit for food,* and *insanitary,* allows common usage interpretation and at the same time provides a basis for specific prohibition of some materials and chemicals in foods. The requirements specified for food production will help to ensure arrival of raw foods at the processing plant in good condition. Such must be the case to comply with the regulation "deemed to be adulterated . . . if it has been prepared, packed, or held under insanitary conditions whereby it may have become contaminated with filth, or whereby it may have been rendered injurious to health . . ." A sanitary food processing plant becomes insanitary when raw products bearing contamination, filth, and so on are accepted into the plant.

Raw Materials

Because there is no possible way to prevent raw materials from bringing in some amount of contamination, and possibly some pathogenic organisms, plant practices to provide sanitation must begin with the receiving area for these raw materials. The establishment of a cleaning area for many raw products permits removal of outside contamination and packaging. The amount and method of cleaning required will vary with the raw product coming in, and will keep outside contamination at a minimum. The product is then moved into storage at the proper temperature to prevent microbial growth.

The activities in the receiving area include inspection; sorting, elimination of damaged, decomposing, or pest-infested materials; and change of the product from shipping containers to fresh, disinfected containers that remain in the plant. In some cases, for example, shellfish, this last activity is prohibited. With this procedure, the processing plant prevents entry into the plant of insects, rodents, and possibly many pathogenic microorganisms. Such prevention is required because the use of many chemicals in the plant for killing such pests may well result in chemical residuals, dead insects, or evidence of rodent presence, which will make the product unfit for human consumption. The prevention of contamination at this point must be absolute because government regulations do not tolerate adulteration.

Prevention of Animal Infestation

When raw products are brought into a plant in the manner just described, with little or no outside contamination, then the next essential to sanitation must be a plant structure that will prevent entrance of contaminants, insects, and rodents. Structural practices making buildings rodent proof are used in much modern construction; however, some measures will assist in preventing rodent infestation in older buildings. First and most obvious, any small openings should be closed off securely by metal plates or by heavy corrosion-resistant screening. Plumbing should be inspected thoroughly to insure that no spaces remain around pipes to permit entrance. Windows and vents should be screened with close-fitting screens, and maintained to prevent removal by wear or accident. Doors must fit securely; and, preferably, each entrance to the plant should be through double doors of a type that will automatically close and remain closed when not in use. These measures to prevent rodent infestation will also be quite effective in preventing entrance of many insects. Screening must be of a smaller gauge for preventing insect entrance, and all closures for all entrances and openings must fit more perfectly. For some insects, such as roaches, chemicals with long residuals painted around all openings are effective to prevent entrance at poorly fitting closures, since both flies and roaches can enter a building through an extremely small opening.

The Dallas City Code specifies the following for insect and rodent control.

Sec. 17-6.6.   INSECT AND RODENT CONTROL.

In order to minimize the presence of insects and rodents, a food products establishment shall:

(1) protect openings to the outside against the entrance of rodents;

(2) equip outside doors that open into an area where food is served or prepared, with tight-fitting, self-closing doors;

(3) maintain windows, except service windows when in use, closed, or equip windows with tight-fitting screening of a size not larger than 16 mesh to the inch; and

(4) equip skylights, transoms, intake and exhaust air ducts, and openings other than doors and windows, that open into an area where food is served or prepared, with tight-fitting glass or screening of a size not larger than 16 mesh.

Perhaps one of the more effective means of preventing rodent and insect infestation of a food processing plant is to keep the premises, inside and out, clean and free of trash, garbage, wastes, and filth which will attract pests to the site. Adequate maintenance of grounds and plant interior in this respect will do much toward avoiding infestation.

Storage Facilities

Although a plant structure in its entirety may not be pest-proof, all plants can provide storage facilities that will prevent contamination of raw or finished products by these animals. Such storage arrangements, together with proper maintenance and cleaning, plus frequent inspection and control programs will help to maintain a plant in satisfactory sanitary condition in regard to such pests.

Many foods require refrigeration during any storage period to prevent spoilage. Such refrigerated spaces, like other storage facilities, must be spacious enough to permit thorough and adequate cleaning and disinfection. Some microorganisms are capable of growing at the ordinary refrigeration temperatures of 50° F (10° C) or below, and their continued growth will produce spoilage or other potential hazards to the product. Continuous and accurate records of the performance of refrigeration equipment are essential to ensure that undetected failures have not occurred and have not damaged foods. All refrigeration and storage facilities must be subjected to the same regular and thorough cleaning and disinfection schedules described for food handling areas.

In food processing or preparation and service, it is frequently necessary that food items, cleaning and disinfection supplies, and other chemicals be stored for various times. Conditions for storage of foods, and for other items are specified in various governmental regulations. The Food Ordinance of the City of Houston, Texas states:

FOOD STORAGE: General.

(a) Food, whether raw or prepared, if removed from the container or package in which it was obtained, shall be stored in a clean covered container except during necessary periods of preparation or service. Container covers shall be impervious and nonabsorbent, except that linens or napkins may be used for lining or covering bread or roll containers. Solid cuts of meat shall be protected by being covered in storage, except that quarters or sides of meat may be hung uncovered on clean sanitized hooks if no food product is stored beneath the meat.

(b) Containers of food shall be stored a minimum of six inches above the floor in the manner that protects the food from splash and other contamination, and that permits easy cleaning of the storage area, except that:

(1) Metal pressurized beverage containers, and cased food packaged in cans, glass or other waterproof containers need not be elevated when the food container is not exposed to floor moisture; and

(2) Containers may be stored on dollies, racks or pallets, provided such equipment is easily moveable.

(c) Food and containers of food shall not be stored under exposed or unprotected sewer lines or water lines, except for automatic fire protection sprinkler heads that may be required by law. The storage of food in toilet rooms or vestibules is prohibited.

(d) Food not subject to further washing or cooking before serving shall be stored in a way that protects it against cross-contamination from food requiring washing or cooking.

(e) Packaged food shall not be stored in contact with water or undrained ice. Wrapped sandwiches shall not be stored in direct contact with ice.

(f) Unless its identity is unmistakable, bulk food such as cooking oil, syrup, salt, sugar or flour not stored in the product container or package in which it was obtained, shall be stored in a container identifying the food by common name.

(g) Bulk packaged food in food warehouses shall be stored at least six inches away from walls and off the floor on approved racks or approved pallets in a way that permits inspection under and behind such storage and in a way that permits the cleaning of the storage area and that protects the food from contamination by splash and other means. Cased food packaged in cans, glass or other waterproof containers need not be elevated when the case of food is not exposed to floor moisture.

The storage and use of poisons or toxic chemicals are covered in separate sections of the Food Ordinance of the City of Houston, Texas.

POISONOUS OR TOXIC MATERIALS: Materials permitted.

(a) Only those poisonous or toxic materials necessary for the maintenance of the establishment, the cleaning and sanitization of equipment and utensils, and the control of insects and rodents shall be present in food service establishments.

(b) Only those pesticides which have been properly registered and approved by appropriate governmental authorities for the purpose of maintaining food service establishments in a sanitary condition shall be used.

Labeling of Materials.

Containers of poisonous or toxic materials shall be prominently and distinctly labeled according to law for easy identification of contents. Each container shall be labeled with the manufacturer's instruction for use.

Storage of Materials.

Poisonous or toxic materials consist of the following three (3) categories:
(a) Insecticides and rodenticides;
(b) Detergents, sanitizers, and related cleaning or drying agents;
(c) Caustics, acids, polishes, and other chemicals.

Each of these categories shall be stored and located to be physically separated from each other. All poisonous or toxic materials shall be stored in cabinets or in similar physically separated compartments or facilities used for no other purpose. To preclude potential contamination, poisonous or toxic materials shall not be stored above food, food equipment, utensils or single-service articles, except that this requirement does not prohibit the convenient availability of detergent or sanitizers at utensil or dishwashing stations.

The Dallas City Code is detailed in itemizing a number of specific foods and conditions (see Appendix 3) and add the following special item storage requirements:

Sec. 17-3.1. POISONOUS SUBSTANCES.

(a) A food products establishment that stores or uses a poisonous substance shall:

(1) keep no poisonous substance on the premises unless the substance is a personal medication or unless the substance is necessary for a pest control, cleaning, or sanitization operation and approved as safe by the director;

(2) store a poisonous substance in a closed container that is labeled in a way that readily identifies its contents; and

(A) in a cabinet used for no other purpose; or

(B) outside an equipment storage area or food preparation, storage, or display area;

(3) store pesticides, cleaning agents, and caustics, each in a separate cabinet so that they are physically separated from each other;

(4) not use or store a poisonous substance in a way that leaves a toxic residue on food-contact surfaces, contaminates food, equipment, or utensils, nor exposes a person in the establishment to a health or safety hazard.

Sec. 17-5.10. HANDLING AND STORAGE OF EQUIPMENT AND UTENSILS.

(b) Storing. A food products establishment that stores food service equipment and utensils shall:

(1) maintain the equipment and utensils in a clean, dry condition, protected from contamination;

(2) store the equipment and utensils in a way that protects them from contamination of a food-contact surface;

(3) store utensils in a noncorrosive, impermeable facility located not fewer than 24 inches above the floor or located in an enclosed storage facility that protects them from contamination;

(4) store flatware in a way that presents the handle to a person's grasp; and

(5) store a food container in an inverted position unless impracticable.

Sec. 17-5.12.   PROHIBITED STORAGE.

A food products establishment shall not store food service equipment or utensils or single service articles;

(1) in toilet rooms or vestibules; or

(2) under exposed sewer or water lines, except automatic fire protection sprinkler heads.

Sec. 17-7.6.   CLEANING EQUIPMENT STORAGE.

A food products establishment shall store and maintain maintenance and cleaning tools, as illustrated by, but not limited to, brooms, mops, and vacuum cleaners, in a manner that does not contaminate food, utensils, equipment, or linens and that facilitates the cleaning of the storage area.

Air-Conditioning

Prevention of air contamination is best accomplished by an adequate air-conditioning system with suitable filtration of incoming air. Although some chemical contamination may not be removed without specialized systems, dust, dirt, smoke, and airborne fungal spores and bacteria will be largely removed by most air-conditioning systems that are cleaned regularly. In the absence of air-conditioning, forced air circulation systems that pass the air through filters will do much to reduce the level of such airborne contamination. The importance of air filtration in practically all locations is easily judged by examination of fan blades of a fan that has run in most rooms for any length of time.

Although air-conditioning is not specified in most ordinances, ventilation requirements usually are detailed; for example in the New York Sanitary Code:

Ventilation. Rooms in which food is prepared or served, or utensils are washed and dressing or locker rooms, toilet rooms, and garbage and refuse storage areas shall be well ventilated by natural or artificial means. Ventilation hoods and devices shall be designed and maintained to prevent grease or condensate from dripping into food or onto food-preparation surfaces. Filters or grease extractors, shall be readily removable or accessible for cleaning or replacement and shall be kept clean. Intake air ducts shall be properly designed and maintained. Ventilation systems shall comply with applicable State and local fire prevention requirements and shall, when vented to the outside air, discharge in such manner as not to create a nuisance. Rooms, areas and equipment from which aerosols, objectionable odors, or noxious fumes or vapors may originate, shall be effectively vented to the outside air and in such a manner as to prevent the creation of a nuisance.

The Dallas City Code serves the same purpose in the following very succinct statement.

A food products establishment shall:

(1) ventilate all rooms in a manner sufficient to keep them free of excessive: heat, steam, condensation, vapors, offensive odors, smoke, and fumes; and

(2) mechanically ventilate to the outside, all rooms from which offensive odors, vapors, or fumes originate.

The Houston Food Ordinance states:

VENTILATION. General. All rooms shall have sufficient ventilation to keep them free of excessive heat, steam, condensation, vapors, obnoxious odors, smoke, and fumes. Ventilation systems shall be installed and operated according to all applicable laws, and when vented to the outside, shall not create an unsightly, harmful or unlawful discharge.

Special Ventilation.

(a) Intake and exhaust air ducts, shall be maintained to prevent the entrance of dust, dirt, and other contaminating materials.

(b) In new or extensively remodeled establishments, all rooms from which obnoxious odors, vapors or fumes originate shall be mechanically vented to the outside.

(c) When such ventilation may result in the deposition of particulate matter or liquids within the ventilation system, ventilation hoods and ventilation equipment shall be equipped with effective, easily removable, easily cleanable filters located adjacent to the intake openings or the intake and exhaust openings of the ventilation system. Such filters shall be cleaned at sufficient frequencies to prevent accumulations.

Water Supply

If a plant obeys the law, in practically any locality, it will not have waterborne contamination problems, because most regulations specifically require the exclusive use of water from an approved source. To assure that no accidental or periodic deficiency occurs in the water quality provided to the plant, contact with the nearest public health laboratory will provide information needed for routine water sampling within the plant. That laboratory will generally test the water samples for safety at no or nominal cost to the plant. The importance of a potable water supply cannot be overemphasized whether or not the water comes into direct contact with the food product. Use of polluted or heavily contaminated water in plant maintenance, equipment and utensil cleaning, or in cooling operations may well result in building up contamination of the plant, its equipment, and the product rather than reducing dangers as intended. Of course, the use of polluted or contaminated water to come into contact with or be added to the food products is not to be tolerated because this will result in adulteration.

Cleaning and Disinfection

Regular schedules of cleaning and disinfection of the plant and equipment are essential for maintaining sanitary conditions. Cleaning routines must include cleaning of all interior surfaces of the building, cleaning of all equipment, removal of all trash, wastes, and garbage, and precautions to avoid leaving chemical residues in locations that could result in food contamination. A clean floor will be of little benefit surrounded by walls, furniture, ceilings, and light fixtures that are dusty, spattered with foods, or otherwise remain filthy. The use of cleaning compounds and disinfectants with residual action in those areas where contact with food does not occur will aid in maintaining sanitary conditions.

Wastes and garbage must be held in containers with close-fitting covers, closed except for deposits. Regular removal of waste and garbage at intervals that do not permit spoilage is necessary. Clutter, shipping or storage containers, wrappings, and so on, should be handled in closed containers as garbage if it is essential in some situations that these be present in operating plant areas. Much of this can be avoided by transfer of products to in-plant containers, as mentioned earlier, which can be immediately returned to the proper plant area.

As in cleaning the plant structure, two essential steps are needed in cleaning equipment and utensils. These are the removal of food particles and disinfection before reuse. Both steps are made easier and may be accomplished more efficiently if equipment and utensils without crevices, joints, indentations, and so on can be used. Smooth, nonporous surfaces with all surfaces easily accessible facilitate the removal of food particles. The use of hot water and a suitable soap or detergent softens particles and permits removal with little scrubbing or brushing. Utensils are most efficiently cleaned by mechanical or automatic washers, simply because water may be used at a higher temperature. For hand washing of utensils, personnel will be unable to use water at a temperature much above 120° F (49° C). Ordinary coliform bacteria will withstand this temperature for more than ten minutes, and more resistant forms will be totally unaffected. Mechanical washers can easily be set for water temperatures between 140° F (60° C) and 170° F (77° C) at which levels non-spore-forming pathogenic microorganisms will be killed, if temperatures are maintained even for a few minutes. Maintenance of temperature levels in a mechanical washer for longer times is more easily accomplished without delaying the total cleaning process. In use of mechanical washers, however, there must be an efficient balance between prerinsing to remove gross particles, and the agitation of the mechanical cleaning process, to assure that all food particles are removed.

In using either hand or mechanical washing, a complete and thorough rinse is essential for removal of all residual detergent or chemicals. For this rinse, an elevated water temperature is also useful to help sanitize or disinfect utensils and to increase the amount of detergent or chemical that can be dissolved for removal. Because, even in hand washing, it is not necessary that workers have their hands in the rinse water, a considerably higher water temperature can and should be used.

For heavily contaminated utensils and equipment and those that could inadvertently have been so contaminated in use, a disinfectant step between wash and rinse may be needed. Cleaning must be a prior step for effectiveness because organic matter such as food particles bind and inactivate many disinfectants, thus preventing the contact of disinfectant with contaminating microorganisms. Disinfection is not a substitute for cleaning; it is an additional process that needs to precede use. Numerous efficient chemical disinfectants are available; however, chlorine and chlorine compounds such as hypochlorite remain among the more popular, perhaps because of the ease of removal by rinse and the reduced likelihood of leaving a residual that is a potential chemical pollutant for foods. As in the use of temperature for sanitization, in the use of chemical disinfectants, sufficient concentration and time of contact are necessary for efficient action. Reduction of concentration requires an increase in time of contact, and such reduction permits disinfection only within certain limits. All chemicals, below specific concentrations, will become ineffective, and should be used only at the recommended and proven levels for the proper time.

Fixed and electrical equipment and appliances present somewhat more difficult problems for cleaning and disinfection; however, it is no less essential that these be regularly and thoroughly cleaned and disinfected. The fact that most such equipment and appliances do remain at room temperature, and are, therefore, potential incubators permitting microbial growth, makes it mandatory that any such objects that come in contact with food be cleaned and disinfected. As with smaller utensils, chemical or heat disinfection is not a substitute for cleaning. The best way to keep such equipment in a satisfactory condition is to wash, regularly and thoroughly, all those surfaces that can be washed, and to follow cleaning with disinfection. For many types of equipment, necessary parts may be disinfected by the use of chemicals. In other instances, this may be difficult and in such cases where surfaces contact foods, there is great danger of microbial contamination and growth. A steam hose is often used for cleaning fixed equipment, and flowing steam may be quite effective for cleaning. In general, however, this cannot be considered to be a method of disinfection, because the time of contact will be too short to reach a temperature on the metal, which will result in killing microorganisms. Only when steam contact with metal is sufficiently long to raise the temperature of the metal to at least 170° F (77° C) for several minutes, will any significant number of microorganisms be killed.

A potential source of contamination in any handling of food products resides in cloths, towels, hot pads, or gloves for handling hot objects in the processing areas. In use, these become soiled and collect food particles, which then serve as a nutrient source for contaminating microorganisms. Such articles should be regularly and routinely changed after use for no more than a few hours; only thoroughly washed and sterilized articles should be used in these areas. A disinfected table top is contaminated as soon as a soiled cloth is used to wipe it.

The Dallas City Code specifies cleaning practices that are acceptable (Appendix 3), as do the 1976 recommendations of the FDA for food service sanitation ordinances. Certain segments of food industries have special requirements, some of which are specified in the chapter dealing with dairy sanitation. All special cleaning practices, as well as those that should be used routinely in working with food at any time, are designed to meet the requirements as set forth in the Food, Drug, and Cosmetic Act as amended, and as set forth in the Consumer Food Act of 1974. The latter law, based on the needs for legislation and monitoring of food processing, distribution, and retailing, expanded the regulation and controls for prevention of food adulteration and foodborne disease outbreaks. These needs can be observed in data included in a report to Congress by the Comptroller General of the United States in 1972, which showed only about thirty-one percent of the food manufacturing industry in complete compliance with the Food, Drug, and Cosmetic Act, and only about another twenty-nine percent having only minor infractions. Obviously then, approximately forty percent of this industry had not operated to protect the public health at that time.

Personnel Practices

The practices just discussed will serve to keep a food processing plant in good sanitary condition only so long as the personnel practice good personal hygiene as well as good sanitation. As discussed in an earlier chapter, the human is not naturally clean, at least from a microbiological standpoint, unless there is considerable effort and energy expended toward that end. Troller gives an excellent discussion of the needs for, and establishment of good programs in personal hygiene for food processing establishments. The human factor in sanitation is among the most important because (a) the human must carry out sanitation practices, and (b) the human, by presence in a specific location where sanitation is being practiced, will negate these efforts unless proper precautions are taken. Every human is a potential carrier of disease-producing and food-spoiling microorganisms in and on the body. If such organisms are in either of these locations, they will also be present on the outer clothing within a very short time after that clothing is put on and will increase in numbers as time passes. The number of microorganisms present on the clothing initially will depend in large part on the personal cleanliness of the individual and the cleanliness of that individual's living quarters. Thus, in any food processing and/ or handling occupation, good personal hygiene at home as well as at work may be important in many ways. Assuming the individual has been educated in good personal hygiene, by the time of reaching work, even though dressed in fresh, clean clothing earlier, there is a good potential that this person's clothing and external skin are contaminated by dust, smoke, grime, bacteria, viruses, and fungi that have been picked up from each person, object, and gust of wind encountered on the way to work. For this reason, in many locations it is required that not only workers, but visitors in certain processing areas wear special protective outer clothing at all times. If the individuals leave such areas, even for short periods, such protective clothing should be replaced by fresh apparel.

Although such protective clothing will help in maintaining sanitation in food processing areas, it will be of little avail if individuals are not required, at all times, to use practices that will maintain a sanitary product through the complete processing phase. This means that workers must be (a) thoroughly trained in hygienic work procedures, (b) regularly supervised and observed to assure that those procedures are followed, and (3) removed from the job for failure to follow adequate procedures or for deliberate violation of specified procedures.

As a necessary measure to assure that the workers can abide by these rules, it is essential that clean, uncluttered, and adequate space is provided in the plant for dressing rooms, toilets, and washrooms. Specific regulations governing such facilities are found most often in local ordinances regulating food processing establishments, but may also be found in state requirements. Words frequently used to specify needs are adequate, clean, equipped with plumbing of sanitary design, well-lighted, accessible, kept provisioned with necessary supplies, and kept in good repair. It might be well for plant managers to add the requirement that each employee use these facilities properly and in a manner that maintains them in a sanitary condition. In many plant situations, hand washing facilities in addition to those in dressing and restrooms should be provided in areas adjacent to, but not actually in, food handling areas. Dressing rooms, restrooms, hand washing facilities, and drinking water fountains must be cleaned and disinfected at the same regular and frequent intervals required for the food handling spaces in the plant. Contamination or infestation in these areas is sure to be spread to the remainder of the plant.

These personnel practices will provide for routine sanitation on the part of the workers. Some additional considerations must occasionally be taken into account. Workers handling foods or entering food handling areas should be required to have a thorough medical examination prior to beginning work; and this examination should be repeated at regular intervals. In many cases there has been much debate about the need for, and the utility of, regular medical examinations, and the requirements for health cards for food industry workers. Troller gives an excellent discussion of the advantages of some sensible medical programs, along with the reasons for not requiring frequent, and expensive testing of personnel that will not give useful information to either the worker or the food industry involved. At best, medical examinations for food workers may detect, and permit the cure of chronic conditions. Certainly, every transient potential pathogen picked up by any person will not be detected, and frequent and expensive examinations of food industry workers for the purpose of issuance of health cards will only add to the cost of foods, and perhaps provide a false sense of security to both the patron of a food establishment, and to the worker who is given the stamp of approval by issuance of the card. Adequate and continual supervision is necessary for the detection of the occasional acute and obvious condition, and provisions should be made in every processing plant to remove such infected workers from food handling areas until a physician has certified that danger of disease transmission is past. All other sanitary practices can be rendered ineffectual by one worker handling a food product when he or she has an obvious common cold infection. On the other hand, an annual, semiannual, or quarterly examination at some other time is not going to detect the presence of this infection if the examination was given last week. Sanitary codes of municipalities and states prohibit the handling of foods by infected persons, and prohibit misconduct in a sanitary sense, by workers. It must be the responsibility of the plant management to provide realistic supervision by a competent individual to ensure that these requirements are followed, and sanitary control is made effective to protect both the food processor and the consumer.

Inspection

Federal, state, and local regulations all require that food industries permit proper authorities to make inspections of the premises where and processes by which food products are prepared, handled, or sold, and that these industries permit the taking of samples for the purpose of testing. Food industries should approve and cooperate with inspections as necessary, and should welcome required sampling as a matter of self-protection. The inspector should be given every opportunity to see all parts of the plant and the process. In this way, he or she may spot potential trouble or shortcomings before real difficulties arise. The U.S. Food and Drug Administration has published useful bulletins dealing with the routines, aims, and utility of inspections for several different food industries. The purposes of inspections, as repeatedly stressed in that literature, is not to cause difficulty nor to harass an industry or plant. The purposes are to be helpful to the industry to produce a product of high quality that will be completely safe for the consumer, and still will be profitable for the industry.

Because the inspector is not a part of the plant routine or a part of the habitual operation of the plant, he or she will see the plant and each step in the process with new eyes, and may notice shortcomings that are so much an ingrained habit as to go unnoticed by regular personnel. The inspector's samples may detect a point of contamination heretofore unsuspected. Although it may on occasion be necessary to condemn an entire lot of product due to sample tests, in the long run the location of the source of trouble will save the industry money.

The most thorough, accurate, and efficient inspection is useless without action of the plant manager and personnel to correct any shortcomings, defects, or potential trouble areas pointed out as a result of the inspection. Any such condition or practice noticed by an inspector is likely to increase in danger to the product unless immediate and thorough steps are taken to correct the situation. As repeatedly pointed out in earlier sections, most disease transmission contaminations become cumulative unless remedial action is taken.

BIBLIOGRAPHY

CITY OF DALLAS.   1977.   Ordinance #15578 Amending Chapter 17, *Food Establishments and Drugs*, of the Dallas City Code. Dallas, TX.

CITY OF HOUSTON.   1978.   *The Food Ordinance*. An Ordinance Amending Article II of chapter 19 of the *Code of Ordinances of the City of Houston* regulating food establishments, requiring the licensing thereof, providing for licensing of food managers; providing a penalty; providing effective dates; providing a severability clause; and declaring an emergency. Houston, TX.

NEW YORK STATE. 1975. *State Sanitary Code*, chapter 1, Part 14. As amended effective 1975. New York.

STATE OF TEXAS. 1962. *Texas Food, Drug, and Cosmetic Act.* Art. 4476-5 V.C.S., Texas Department of Health. Austin, TX.

TEXAS STATE DEPARTMENT OF HEALTH. 1978. *Rules on Food Service Sanitation.* Division of Food and Drugs. Austin, TX.

TROLLER, J. A. 1983. *Sanitation in food processing.* New York: Academic Press.

U.S. DEPARTMENT OF HEALTH, EDUCATION, AND WELFARE. 1974. *Federal Food, Drug, and Cosmetic Act,* as amended. Washington, DC: U.S. Government Printing Office.

U.S. DEPARTMENT OF HEALTH, EDUCATION, AND WELFARE. 1976. *Food service sanitation manual.* Public Health Service, FDA. DHEW Publication no. (FDA) 78-2081. Washington, DC.

9

Dairy Food Sanitation

Foodservice operations have little control over the sanitary aspects of many dairy products, because these products arrive at the foodservice site in pre-packaged, pre-sanitized units. Most ordinances dealing with milk or milk products at the foodservice level simply state the requirements as to sources of these foods and specify storage and service conditions. For example, in the Food Ordinance of the City of Houston, Texas, other than requiring that all food be obtained from sources complying with all laws relating to food and food labeling, there are the following special requirements for dairy foods. "(a) Fluid milk and fluid-milk products used or served shall be pasteurized and shall meet the Grade A quality standards as established by applicable laws. Dry milk and dry-milk products shall be made from pasteurized milk and milk products."

In reference to food preparation, the ordinance discusses dry milk: "Dry milk and dry milk products. Reconstituted dry milk and dry milk products may be used in instant desserts and whipped products, or for cooking and baking purposes."

The ordinance states the following requirements for food display and service:

Potentially hazardous foods. Potentially hazardous food shall be kept at an internal temperature of 45° F (7° C) or lower or at a temperature of 140° F (60° C) or above during display and service, except that . . . .

Milk and cream dispensing.

(a) Milk and milk products for drinking purposes shall be provided to the consumer in an unopened, commercially filled package not exceeding one pint in capacity, or drawn from a commercially filled container stored in a mechanically refrigerated bulk milk dispenser. Where it is necessary to provide individual servings under special institutional circumstances, milk or milk products may be poured from a commercially filled container provided such a procedure is authorized by the Health Officer. Where a bulk dispenser for milk and milk products is not available and portions of less than one-half pint are required for mixed drinks, cereal, or dessert service, milk and milk products may be poured from a commercially filled container.

(b) Cream or half and half shall be provided in an individual service container, protected pour-type pitcher, or drawn from a refrigerated dispenser designed for such service.

Nondairy product dispensing.

Nondairy creaming or whitening agents shall be provided in an individual service container, protected pour-type pitcher, or drawn from a refrigerated dispenser designed for such service.

State laws generally do not deal with dairy products except in the area of foodservice establishments. For example, the Texas Pure Food and Drug Law does not deal with dairy food requirements; however the Texas Rules on Food Service Sanitation state:

(2) Special Requirements.

(A) Fluid milk and fluid milk products used or served shall be pasteurized and shall meet the Grade A quality standards as established by law. Dry milk and dry milk products shall be made from pasteurized milk and milk products.

(2) Milk and Cream Dispensing.

(A) Milk and milk products for drinking purposes shall be provided to the consumer in an unopened, commercially filled package not exceeding one pint in capacity, or drawn from a commercially filled container stored in a mechanically refrigerated bulk milk dispenser. Where it is necessary to provide individual servings under special institutional circumstances, milk and milk products may be poured from a commercially filled container provided such a procedure is authorized by the regulatory authority. Where a bulk dispenser for milk and milk products is not available and portions of less than one-half pint are required for mixed drinks, cereal, or dessert service, milk and milk products may be poured from a commercially filled container.

(B) Cream or half and half shall be provided in an individual service container, protected pour-type pitcher, or drawn from a refrigerated dispenser designed for such service.

(3) Nondairy product dispensing. Nondairy creaming or whitening agents shall be provided in an individual service container, protected pour-type pitcher, or drawn from a refrigerated dispenser designed for such service.

Obviously, the City of Houston has taken the state requirements in almost a word-for-word fashion. The same is also true for cities in other states, and in most states, the requirements for foodservice are specifically stated. For example, the Rules and Regulations for Food Service for the State of Georgia state:

(c) all food in the food-service establishment shall be wholesome and free from spoilage, adulteration, and misbranding.

(2) Milk and Milk Products:

(a) all milk and milk products, including fluid milk, other fluid dairy products and manufactured milk products, shall meet the standards of quality established for such products by applicable State and local laws and regulations;

(b) only pasteurized fluid milk and fluid-milk products shall be used or served. Dry milk and milk products may be reconstituted in the establishment if used for cooking purposes only;

(c) all milk and fluid-milk products for drinking purposes shall be purchased and served in the original, individual container in which they were packaged at the milk plant, or shall be served from an approved bulk milk dispenser: Provided, that cream, whipped cream or half and half, which is to be consumed on the premises, may be served from a dispenser approved by the health authority for such service, and for mixed drinks requiring less than one-half pint of milk, milk may be poured from one-quart or one-half gallon containers packaged at a milk plant.

These regulations uniformly follow the recommendations embodied in the Grade A Pasteurized Milk Ordinance as promulgated by the U.S. Department of Health, Education, and Welfare, Food and Drug Administration, as it has been modified through the years as technology changed. The FDA does not have actual regulatory authority over the quality and sanitation of milk and milk products within any given state. Here the FDA can recommend, but not require, certain standards of practice in the dairy industry. In interstate movement of dairy products, however, the FDA does have jurisdiction, and the dairy foods are subject to the requirement of the Federal Food, Drug, and Cosmetic Act. State and local authorities have been very successful in promulgating and achieving high sanitary standards in the dairy industry. These standards and this level of sanitation have been achieved by cooperation and assistance from the federal, state, local governments (including both health and agriculture departments), dairy producers, plant operators, equipment manufacturers, and many educational and research institutions. As stated by Richard D. Vaughan in the Foreword of the 1965, Pasteurized Milk Ordinance (U.S. PHS Publication No. 229), "The responsibility for insuring the ready availability and safety of milk and milk products is not confined to an individual community or a State, or to the Federal Government—it is the concern of the entire Nation. With the continued cooperation of all interested groups, both Government and industry, engaged in the sanitary control of milk and milk products, such responsibility can be accepted with confidence." Although technology may have changed since this was written, the truth of this statement has not changed.

Throughout the history of food sanitation, the dairy industry has generally been among the leaders with respect to development of sanitation practices and standards. Although the recognition that dairy products could be and often were sources for infectious diseases influenced this leadership, the recognition that good sanitation practices would be financially rewarding to the industry also influenced the industry leaders. If the producers of dairy products minimize source contamination, suppress microbial multiplication, apply pasteurization, and prevent recontamination during production and processing, then good quality milk and milk products will reach the foodservice operation, or the consumer.

In recent years, the FDA has modified and promulgated the Grade A Pasteurized Milk Ordinance and Code and set the requirements as shown in Table 9.1 for milk of various types to be sold as Grade A products as required by so many local ordinances, as seen earlier.

These bacterial counts can be achieved only by rigid maintenance of temperatures at all stages of production and shipment. The recommendations of the Standard Ordinance and Code are that raw milk be maintained at 50° F (10° C) or less until processed, and that it should be cooled after pasteurization to 45° F (7° C) and then maintained at this temperature or below. Producers, plant operators, and retail establishments attempt to keep milk at a temperature of 41° F (5° C) to increase shelf life of the product until and after it reaches the consumer.

It is apparent from the local ordinances applying to foodservice, and from the bacterial counts recommended as maximum allowable, that the primary aim of all is to keep milk from being contaminated, and to prevent growth of bacteria that will be present from the time of production. If raw milk is allowed to have 100,000 bacteria/ml at the source (dairy producer), this means that milk, as it comes from the cow, is expected to have sizeable bacterial populations, and that production (i.e., any handling) will increase that population. The fact that

Table 9.1.   Requirements For Some Grade A Dairy Products

Product	Standard Plate Count/ml	Coliforms/ml
Raw milk, source	100,000	NA
Raw milk, at plant	300,000	NA
Pasteurized milk & products	20,000	10
Condensed milk	30,000	10
Nonfat dry milk	30,000	10

there is no allowed maximum for coliforms at this stage means that some may very well be present, but are not of great concern now, as long as the total population of all bacteria is not above 100,000/ml, because the milk is still to be pasteurized. The lack of a coliform standard for raw milk at the plant is considered in the same way, although the total bacterial population is now allowed to be greater. There is, however, still a maximum limit set for the total bacterial count of raw milk at the plant (after milk from several producers has been mixed, or "comingled"). The reason for this maximum, although higher than at the producer is (a) to prevent gross mishandling or contamination during transport to the plant which might otherwise go undetected; (b) to prevent standing or transport at the wrong temperature thereby allowing massive growth of bacteria; and (c) to prevent the addition of unfit milk to lots of good milk (thereby diluting the total contamination and making it undetectable) by unscrupulous operators. The permitted tripling of total bacterial numbers at this stage is simply realistic as to the allowable multiplication of bacteria present in the raw milk at production, even when properly handled, transported, and maintained at the correct temperature.

The prohibition in most local ordinances of service of milk to a customer from any container other than the one used at the processing plant is a recognition that any handling of milk or milk products after pasteurization presents an opportunity for contamination. As with any food, each handling step that is removed reduces final contamination levels. Allowed serving of cream, milk for cereal, and milk for use in mixed drinks from large containers rather than in individual containers is realistic in that packaging of these small amounts would increase the cost to too great an extent. These regulations make the work of the foodservice establishment easier, and without doubt, result in much less contamination of milk and cream with pathogenic organisms. Only a very small percentage of foodborne disease is transmitted by milk or milk products from foodservice establishments. Most foodborne disease is due to contamination of other types of food, and it is probable that this fact is in large part due to the practice of reducing handling of milk after pasteurized containers have been opened.

Milk, both fresh and processed, in the form of condensed or dry milk, is an excellent medium for the growth of a large number of different kinds of bacteria. Milk fat, in the form of heavy cream or butter fat, is not as good a culture medium as whole milk; however, these products will support the growth of many microorganisms. Cheeses, like other high-fat-content milk products, while not quite as good a growth medium for bacteria, will support the growth of some organisms capable of causing disease in humans, including both bacteria and fungi.

Of all of these types of dairy products, the one most likely to have received excessive contamination is dry milk. Dried milk should be made from pasteurized milk and processed with care to prevent excessive contamination by extraneous microorganisms. The process of drying, however, is not sufficiently stressful to kill most microorganisms, and dried milk and milk products should be handled most carefully in foodservice establishments. Most local ordinances prohibit the service of reconstituted dried milk or dried milk products directly to the consumer, and these products are permitted to be used only for cooking. Even in this use, however, excessive contamination may cause difficulty in some cases, and great care should be taken in storage and use of such products to avoid using contaminated utensils in measuring or dispensing in the process of cooking. Once reconstituted, this milk is equally as good as fresh, whole milk as a microbial culture medium, and must be refrigerated and handled accordingly. A thorough discussion and consideration of the proper techniques for handling milk and milk products in a foodservice establishment is given by Longree and Blaker in *Sanitary Techniques in Foodservice.*

Although the FDA does not have the authority to directly enforce the Grade A Milk Ordinance, it does have the authority and duty to enforce the Filled Milk Act. This regulation was passed by the U.S. Congress to regulate the production and sale of milk products for interstate or foreign commerce that are made by the addition of other materials. The Act defines *filled milk* as:

any milk, cream, or skimmed milk, whether or not condensed, evaporated, concentrated, powdered, dried, or desiccated, to which has been added, or which has been blended or compounded with, any fat or oil other than milk fat, so that the resulting product is in imitation or semblance of milk, cream, or skimmed milk, whether or not condensed, evaporated, concentrated, powdered, dried, or desiccated. This definition shall not include any distinctive proprietary food compound not readily mistaken in taste for milk or cream or for evaporated, condensed, or powdered milk, or cream where such compound (1) is prepared and designed for feeding infants and young children and customarily used on the order of a physician; (2) is packed in individual cans containing not more than sixteen and one-half ounces and bearing a label in bold type that the content is to be used only for said purpose; (3) is shipped in interstate or foreign commerce exclusively to physicians, wholesale and retail druggists, orphan asylums, child-welfare associations, hospitals, and similar institutions and generally disposed of by them.

The FDA is also responsible for the enforcement of the Federal Import Milk Act, which requires that importation of milk or cream requires a valid permit from the Secretary of Health and Human Services of the United States. This act specifically details the conditions and circumstances that make milk or cream unfit for importation into the United States. One of these circumstances states that these products are unfit for importation "when all cows producing such milk or cream are not healthy and a physical examination of all such cows has not been made within one year previous to such milk being offered for importation; . . . ." This act also authorizes the secretary to use his or her judgment" for the inspection of milk, cream, cows, barns, and other facilities used in the handling of milk and/or cream and the handling, keeping, transporting, and importing of milk and/or cream: . . . ." This act also includes a section stating "Nothing in sections 141 to 149 of this title is intended nor shall be construed to affect the powers of any State, or any political subdivision thereof, to regulate the shipment of milk or cream into, or the handling, sale, or other disposition of milk or cream in, such State or political subdivision after the milk and/or cream shall have been lawfully imported under the provisions of the said sections."

MILK-TRANSMITTED DISEASES

Historically, streptococcus infections, diphtheria, and tuberculosis were the major diseases likely to be transmitted by milk. Salmonella and other gastroenteritis infections have been occasionally transmitted by milk throughout history, but these generally were not as common or at least were not as life-threatening as those above, and were generally thought to be preventable by good hygienic practices in handling milk and dairy products. All of these infectious disease organisms were rather routinely handled by the process of pasteurization, and it is this process that has reduced the transmission of diseases by milk from fairly regular to relatively rare. In processing plants, and in the use of raw milk, however, there continue to be considerable outbreaks of infectious diseases as a result of consuming milk or milk products. Perhaps the most notable example of this kind of transmission occurred in 1985 in the Midwest as a result of difficulty in a processing plant. In that episode, over 5,000 cases of infection with *Salmonella typhimurium* occurred as a result of consumption of processed two percent fat milk from one dairy plant. This outbreak produced the largest number of culture-confirmed cases ever associated with an outbreak of salmonellosis in the United States. Salmonella strains are sometimes found in dairy cattle and in raw milk; however, pasteurization kills these organisms. In the milkborne outbreak reported here, the milk was obviously either inadequately pasteurized, or was contaminated after pasteurization.

In recent years, another intestinal-infecting bacterium, *Campylobacterium jejuni*, has become of concern as a food-transmitted organism, and this also holds true in the case of milkborne disease. Since 1980 there have been several cases of campylobacteriosis as a result of consumption of raw milk. In several U.S. states, it is permissible to sell raw milk directly to consumers; however, it is not permissible for foodservice establishments to use or serve raw milk, according to most local ordinances. The cases of this infection reported in recent years have all resulted from consuming raw milk either at home or at the dairy.

Yersiniosis is a bacterial infection caused by *Yersinia enterocolitica*, which has in recent years been reported to have been transmitted by milk or milk products. In 1982, for example, 172 cases were diagnosed by culture in Tennessee, Arkansas, and Mississippi. Follow-up investigation confirmed that these cases resulted from drinking pasteurized milk from one specific plant. Further investigation could not locate any breakdown in the pasteurizing process, and because the organism is commonly found in raw milk, it is assumed that the contamination of one lot of milk occurred after pasteurization. Subsequent lots of milk tested from this plant were found not to be contaminated with the pathogenic organism. In other studies, it has been found that very large numbers of viable *Yersinia enterocolitica* may, indeed, survive pasteurization, which points up the necessity to utilize good grade raw milk in any pasteurization process. This organism has also been isolated from pasteurized chocolate milk, tofu, and cheeses.

A continuing infectious disease problem in dairy products and in meats is the occasional occurrence of cases of brucellosis. Brucellosis is a febrile infection caused by one of three species of *Brucella* organisms. Its occurrence has been greatly reduced in the United States by immunization of cattle, pasteurization of milk and milk products, and destruction of infected animals. Yet the infection appears occasionally in spite of the rather extensive efforts at control. When the infection was recognized in the United States in recent years, it has generally been found to have resulted from consumption of cheese made from unpasteurized milk. Usually these outbreaks have consisted of only one or a few cases, but in a recent episode, the number of cases was considerably higher.

Listeriosis is a bacterial disease causing meningitis and sepsis, especially in pregnant women and their infants. Since 1984 there have been outbreaks of listeriosis in the United States as a result of consumption of cheese foods, and otherwise there have been outbreaks associated with consumption of cabbage and of pasteurized milk. In the first six months of 1985, there were eighty-six cases of infection by *Listeria monocytogenes* infection in California associated with consumption of a Mexican-style cheese. Of these cases, there were twenty-nine deaths. In the study that followed, the organism was isolated from cheese samples, and the manufacturer recalled the products from the market in seventeen states. Although the milk from which the cheese was made had been pasteurized, there is some evidence that the organism is able to survive ordinary pasteurization temperatures.

These outbreaks of infectious diseases resulted from failures in the processing plants for dairy products, or in the raw milk cases, from the lack of processing. However, it is likely that these instances of infectious disease transmission were noted primarily because of the numbers of individuals involved, and that there were other instances of transmission of one or two cases of infection by milk or milk products that were not detected and reported. These failures in processing, however, point out the fact that milk and milk products are good culture media for bacteria, and without regular and systematic care and sanitary practice, infectious organisms may be transmitted to smaller numbers of individuals in foodservice operations as well as in processing. Such cases have occurred in recent years in small numbers of infections produced by enterotoxigenic strains of *Escherichia coli* in imported soft cheeses, and of *Salmonella* sp. infections from ice cream and raw milk. The dairy industry and the consumer must constantly remain alert to exercise good sanitary practices when dealing with these products that provide such a good culture environment for so many microorganisms.

In the processing of milk and milk products, there is danger of failure of the system and contamination of the products with chemicals as well as with microorganisms. In 1986, there was at least one instance reported where a breakdown in a dairy plant resulted in ammonia contamination of the milk produced, with twenty children affected by the chemical. Early diagnosis, surveillance, and investigation identified 268 schools that had received 520 cartons of contaminated milk. There were no deaths attributed to the incident.

These reports emphasize the importance of sanitation and all hygienic practices in the dairy processing plant. Because foodservice operations are in most cases limited to service of milk and milk products in the pasteurized containers produced in these plants, any breakdown in the plant may well show up at the foodservice point. Some aspects of needed sanitation at these plants are considered in the following pages.

DAIRY PROCESSING PLANTS

Sanitation Programs

The sanitation program of any food operation is only as good as the management and the delivery of that program. It is therefore essential that management be trained in what is needed, and in how to go about its production. This requires management involvement in planning programs and training, and in supervision of all phases of carrying out the program. Any plant program includes cleaning, sanitation, quality control, and supervision.

An obvious first step in any sanitation program is to convince personnel, both management and those carrying out the program, that there is a difference between cleaning and sanitation. No sanitation program that is not preceded by cleaning will be successful. It is first necessary that soil be removed from an environment before that environment can be sanitized. This is usually done by a pre-rinse and in the case of a dairy plant, it is necessary that considerable effort be expended in the cleaning process because of the nature of soils that will be present in and on equipment to be cleaned. Dairy plant soils are most likely to consist of protein, fats, sugars, and salts (especially calcium and phosphorous salts). The nature of these soils is such that the fats and some of the proteins are likely not to be water soluble, and that some of the salts may be water insoluble (see Table 9.2). This then requires that the proper equipment, process, and cleaning compounds be selected for cleaning. Because different cleaners are required,

Table 9.2.   Soil Characteristics

Component on Surface	Solubility Characteristics	Ease of Removal	Changes Induced by Heating Soiled Surface
Sugar	Water soluble	Easy	Caramelization, more difficult to clean
Fat	Water insoluble, alkali soluble	Difficult	Polymerization, more difficult to clean
Protein	Walter insoluble, alkali soluble, slightly acid soluble	Very difficult	Denaturation, much more difficult to clean
Salts			
Monovalent	Water soluble, acid soluble	Easy to Difficult	Generally not significant
Polyvalent ($CaPO_4$)	Water insoluble, acid soluble		Interactions with other constituents, more difficult to clean

it is also necessary that the type of water provided will allow the cleaners to work at maximum efficiency. Waters are classified as hard to soft with grades between, and chemical cleaners of different types behave differently in water of various hardnesses (i.e., different concentrations of calcium and magnesium salts). Generally, water with less than 75 mg/l of $CaCO_3$ is considered soft, and above this is considered hard. If water of the correct type is not readily available from private or municipal sources, then the available water must be treated to make it acceptable. Equipment in processing plants is designed to be cleaned in place (CIP), or in some cases to be dismantled and separate parts cleaned and then reassembled (COP). Cleaning is most easily done if parts can be separated, and all surfaces made available to the cleaning process, whether physical or chemical. For CIP operations it is essential that the proper concentration of the chemical, the proper temperature for maximum chemical effect, the proper time for optimum chemical action, and the proper force of application of the cleaning process be utilized in each instance. Any mistake in any of these factors will cause a failure of the cleaning operation.

Guidelines for selection of cleaning procedures, chemicals, times, and temperatures can be found in many writings, including the most recent revisions of the *Grade A Pasteurized Milk Ordinance* recommendations from the Public Health Service. As technology is improved, improved cleaning compounds are made available for many purposes, including use in the dairy processing industry. Cleaning compounds selected should be able to adequately soften the water; quickly dissolvable; non-corrosive; non-toxic; economical; stable; non-caking; and relatively easy to measure to achieve the desired results. No single chemical will suit all these requirements for all jobs to be done in the dairy industry, and choices must be made in individual situations.

Equipment standards for the dairy industry are set by the Standards Committee (made up of representatives from the Association of Milk, Food and Environmental Sanitarians, American Public Health Service Association, and the Association of Dairy and Food Manufacturers, and the proper division of the FDA). Any equipment chosen for a plant should be designed to be easily cleaned. This is most often accomplished by using good welding on stainless steel parts for most of the plant. Sharp corners, cracks, threads, and rough surfaces should be eliminated, whether internal or external in any equipment used. Unless all parts of the equipment are accessible internally, it must be made so that it can be disassembled for inspection and/or cleaning. Contact surfaces of all equipment should be chemically inert so as not to react with products or chemicals used in cleaning or sanitizing. All equipment should be located within the plant so that there is adequate ventilation and lighting available to conform to the requirements of the *Grade A Milk Ordinance*. Again, standards change as technology changes, and updates from the proper authorities are regularly needed.

Sanitization procedures and chemicals are specified in the recommendations contained in the *Grade A Pasteurized Milk Ordinance*, and these can be found in the publications of the FDA. Sanitizers must not react with surfaces of equipment, and also must be removable by rinse or cleansing in order that no harmful residues be retained by the equipment to be passed on to the consumer of the dairy product.

Sanitization can be accomplished by several agents. Heat, radiation, and chemicals will all sanitize if applied in the proper manner and for the proper time. Heat may be applied as steam, hot water, or hot air. Of these three, steam, because of its temperature, is most efficient in terms of time. Hot water and hot air vary in efficiency depending upon the degree of temperature used. Ultraviolet radiation is very efficient in sanitizing if it can reach the contaminating cells. Ultraviolet light does not penetrate glass, and does not travel great distances effectively enough for sanitizing. There are a large number of chemical agents that can accomplish sanitization. In many situations, chlorine compounds are most desirable as chemical sanitizing agents, because most of the time these can be used in a safe, non-corrosive form and will not leave an undesirable residue. Other chemical sanitizing agents may be preferable for various reasons in some cases; such agents including quaternary ammonium compounds, iodine compounds, acid-anionic agents, alkalis, acids, and bromine-chlorine combinations. In choosing the sanitizing agent, it is necessary that the user consider the specific task at hand, the cost of sanitization, the time required for sanitization, and the overall advantages and disadvantages of the chemical in question.

Quality control, or a regular testing program is essential to all dairy plant operations. Regular testing programs are carried out by most local health authorities to assure the biological safety and proper composition of most dairy foods that are commercially available within that jurisdiction. The *Grade A Pasteurized Milk Ordinance* recommends

During any consecutive 6 months, at least four samples of raw milk for pasteurization shall be taken from each producer and four samples of raw milk for pasteurization shall be taken from each milk plant after receipt of the milk by the milk plant and prior to pasteurization. In addition, during any consecutive 6 months, at least four samples of pasteurized milk and at least four samples of each milk product defined in this Ordinance shall be taken from every milk plant. Samples of milk and milk products shall be taken while in possession of the producer or distributor at any time prior to final delivery.

The ordinance also states:

Samples of milk and milk products from dairy retail stores, food service establishments, grocery stores, and other places where milk and milk products are sold shall be examined periodically as determined by the health authority; and the results of such examination shall be used to determine compliance. . . . Proprietors of such establishments shall furnish the health authority, upon his request, with the names of all distributors from whom milk or milk products are obtained.

These testing programs are specified by local ordinance, and plant management will eventually receive the results of such testing, particularly if the results of the test do not conform to the standards set by the ordinance. It is much to the advantage of management, however, to have in place a functioning laboratory system that will permit more frequent testing of the products and processes with which the plant is concerned. If tests are carried out efficiently and often, trouble spots are located before they become problems, and corrections are made before defective products reach the consumers. The proper testing procedures are regularly updated in the *Standard Methods for the Examination of Dairy Products* (the most recent edition of which was the 15th), a publication of methods to be used for quality control published by the American Public Health Association, Washington, DC.

BIBLIOGRAPHY

CITY OF HOUSTON, TEXAS.   1978.   *The food ordinance*. An ordinance amending article 2, chapter 19 of the *Code of Ordinances of the City of Houston, Texas*.

HUI, Y. H.   1979.   *United States food laws, regulations, and standards*. New York: John Wiley and Sons.

LONGREE, K., and BLAKER, G. G.   1982.   *Sanitary techniques in foodservice*. 2nd ed. New York: John Wiley and Sons.

RICHARDSON, G. H.   1985.   *Standard methods for the examination of dairy products,* 15th ed. Washington, DC: American Public Health Association.

STATE OF GEORGIA.   1980.   *Rules and regulations. Food service*. Chapter 290-5-14. Atlanta, GA.

TEXAS STATE DEPARTMENT OF HEALTH.   1978.   *Rules on food service sanitation*. Austin, TX: Division of Food and Drugs.

U.S. DEPARTMENT OF HEALTH, EDUCATION, AND WELFARE, FDA.   1965.   Recommendations of the United States Public Health Service. *Grade A Pasteurized Milk Ordinance*. Washington, DC.

U.S. DEPARTMENT OF HEALTH, EDUCATION AND WELFARE, FDA. 1971. *Grade A condensed and dry milk products. A recommended sanitation ordinance for condensed and dry milk products used in Grade A milk products.* Washington, DC.

U.S. DEPARTMENT OF HEALTH, EDUCATION AND WELFARE, FDA. 1965. *Fabrication of single-service containers and closures for milk and milk products. Guide for Sanitation Standards.* Washington, DC. (May be revised.)

U.S. DEPARTMENT OF HEALTH, EDUCATION AND WELFARE, FDA. 1966. *Methods of making sanitation ratings of milksheds.* Washington, DC. (May be revised.)

U.S. DEPARTMENT OF HEALTH, EDUCATION AND WELFARE, FDA. 1975. *Milk laboratories approved by federal and state agencies.* Washington, DC. (May be revised.)

U.S. DEPARTMENT OF HEALTH, EDUCATION AND WELFARE, FDA. 1975. *Procedures governing the cooperative state-public health service program for certification of interstate milk shippers.* Washington, DC.

U.S. GOVERNMENT PRINTING OFFICE. 1981. *Laws enforced by the U.S. Food and Drug Administration.* Washington, DC.

10

Food Processing Plant Organization for Sanitation

Food processing is a broad and complex manufacturing activity. It includes food canning, freezing, drying, irradiation, and in some cases preparation, including packaging in individual portions. It is stretched from the plant preparing and packaging cookies and chips for a vending machine, to preparation of lunches for the mobile vending operation for construction jobs, to seafood cleaning and preparation, to vegetable cleaning and preparation, to animal slaughter and dressing, to canning plants, freezing plants, drying plants, and currently to irradiation plants. After years of hearings, considerations, and reviews, the FDA has recently permitted expansion of the use of irradiation for preservation of food supplies. Irradiation may now be used in many instances for extension of the shelf life of foods, and to protect it from insects without creating health hazards to consumers.

For many years, food processing was primarily a matter of producing such food basics as flours, meals, grain products such as cereals, and dried foods because these were the major foods purchased by the consuming public for preparation and use in the home. This era was followed by a period when the major addition to the processed food supply was canned food. Canning today remains one of the major methods for preservation of foods worldwide. The process has changed much since early days in that the types of containers used have been improved to lengthen shelf life of foods, and to present less danger to the consumer from chemical contamination of the foods contained therein.

The canning plant remains one of the specific areas that is of vital concern to the safety of foods for the general public. It is essentially an extension of the home kitchen in that the canner does some of the work that would otherwise have been done in the home kitchen. This means that the canning plant must provide a product of highest quality and purity in order that the health and well-being of the public may be protected. For the canner, this means that an absolute necessity for the plant is a basic, efficient, and effective sanitation program. All such programs consist of three basic parts: planning (including organization), execution, and control. Properly combining these three parts of the sanitation program in a canning plant result in a product of highest quality and purity, and guarantee that the products will remain popular with the consuming public.

CHAIN OF COMMAND IN THE SANITATION PROGRAM

Personnel involved in the sanitation program will include management, sanitarians, quality control officers, and general supervisors. It is the degree of involvement, and the place of involvement of these individuals in the chain of command that operates the plant that determines the eventual outcome of any sanitation program. All management of a food canning plant, from the very top through all junior executives, have been given the legal and moral responsibility for everything that happens within that manufacturing plant. The Food, Drug, and Cosmetic Act, without specifically pointing to any one person, makes top management responsible for the sanitary quality of foods produced in that plant for sale to the public. It then is incumbent on top management to delegate the authority for junior management to carry out the efficient program. To avoid "buck passing," one specific person must be assigned the responsibility to see that the program is accomplished. This person is frequently a vice-president who only reports directly to top management, and does not report to, and is not responsible to, in any way, the person whose interest is in increasing production and sales, or the person whose interest is in keeping the machinery operating. It is practical that the sanitarian in charge of the program report to the plant superintendent, but it is necessary that the program be operable without direction from the plant superintendent. A workable chain of command to accomplish this state might be as follows:

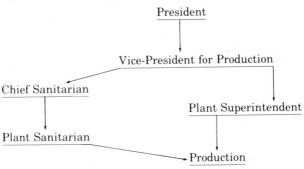

In other words, the sanitarians responsible for the plant sanitation program must be independent of all other programs within the plant, although they will help to implement and actually carry out those programs, and must furnish information that will affect the outcome of those programs. It is the top management of the company who must sign the agreements for inspection and certification plans with the proper government agencies, and who actually agree to follow the Good Manufacturing Practice regulations. But, it is the sanitarian who must implement the practices that accomplish the sanitation or the Good Manufacturing Practice program. The authority for doing so comes to the sanitarian from top management.

The role, position and independence of the sanitarian within any organization must be understood and respected by all parties concerned, including the sanitarian, within any organization. Because this is necessary, one of the most critical jobs of top management is the writing of the job description of the sanitarian, and the clear definition of responsibilities that go with that position. The authority position of the sanitarian must be at least equal to that of department heads such as production, purchasing, engineering, and so on, and all department heads must operate with clear understandings of the interrelationships among the departments.

COMPONENTS OF THE SANITATION PROGRAM

The most effective and least expensive means of accomplishing a good sanitation program is to establish routine procedures for preventive sanitation, and thus eliminate the need for after-the-fact correction of deficiencies. One preventive system of control that is widely used in the United States is the Hazard Analysis and Critical Control Points (HACCP). Such a system is composed of a risk analysis of different foods, a determination of the critical control points of processing different foods, an assessment of effects on different finished products, and a system for recalling finished products from the market when it is learned that they may be hazardous. Such hazards may be microbiological, chemical, physical, or nutritional in nature, and all must be considered in dealing with the HACCP program within the food plant. These potential hazards are all dealt with in the Food, Drug, and Cosmetic Act of the federal government in the section that not only prohibits contamination but also states that "a food shall be deemed to be adulterated if it has been prepared, packed, or held under insanitary conditions whereby it may have become contaminated with filth, or whereby it may have been rendered injurious to health." This section of the law is the basis for the "umbrella" regulations included in the

Good Manufacturing Practices that cover the entire food industry. In general, these GMPs deal with the components of sanitation programs that include housekeeping, animal infestation, insanitary or unsatisfactory equipment, poor personnel practices or hygiene, improper waste handling, and unapproved water supply. Each of these components is affected by, and affects the others, and all affect the sanitary aspects of the food industries concerned.

Housekeeping

Housekeeping, for a food processing plant, designates conditions that determine order and cleanliness within and outside the plant. Housekeeping is dependent in large part on the proper planning and design of the plant. Perhaps the single most important aspect of plant design from the housekeeping standpoint is the design and execution of proper and adequate storage. Such storage must include storage areas for raw materials, in-process materials, finished products, clean utensils, sanitized utensils, waste and rubbish, cleaning agents, disinfectants, sanitizers, pesticides, testing materials, work clothing (both clean and soiled), street clothing, packaging materials, shipping cartons, seasonal equipment, tools and replacement parts, and so on. Responsibility for housekeeping must be assigned to an individual, who in turn must have teams of assistants (official or unofficial) in different areas. However, maintenance of good housekeeping routines and practices must be included in the responsibilities of the foremen, maintenance personnel, machinery operators, all workers, and all management personnel; in other words, it is the responsibility of all plant personnel.

For any food processing plant, the basis for the housekeeping program and the sanitation program remains the regulations covering good manufacturing practice (GMPs) promulgated by the FDA.

128.2. Current good manufacturing practice (sanitation).
The criteria in § § 128.3–128.8 shall apply in determining whether the facilities, methods, practices, and controls used in the manufacture, processing, packing, or holding of food are in conformance with or are operated or administered in conformity with good manufacturing practices to assure that food for human consumption is safe and has been prepared, packed, and held under sanitary conditions.

These GMPs specifically delineate conditions that must be maintained in and around a food processing plant. The regulations include:

§ 128.3. Plant and Grounds.

(a) *Grounds.* The grounds about a food plant under the control of the operator shall be free from conditions which may result in the contamination of food including, but not limited to, the following:

(1) Improperly stored equipment, litter, waste, refuse, and uncut weeds or grass within the immediate vicinity of the plant buildings or structures that may constitute an attractant, breeding place, or harborage for rodents, insects, and other pests.

(2) Excessively dusty roads, yards, or parking lots that may constitute a source of contamination in areas where food is exposed.

(3) Inadequately drained areas that may contribute contamination to food products through seepage or foot-borne filth and by providing a breeding place for insects or microorganisms.

If the plant grounds are bordered by grounds not under the operator's control of the kind described in subparagraphs (1) through (3) of this paragraph, care must be exercised in the plant by inspection, extermination, or other means to effect exclusion of pests, dirt, and other filth that may be a source of food contamination.

(b) *Plant construction and design.* Plant buildings and structures shall be suitable in size, construction, and design to facilitate maintenance and sanitary operations for food-processing purposes. The plant and facilities shall:

(1) Provide sufficient space for such placement of equipment and storage of materials as is necessary for sanitary operations and production of safe food. Floors, walls, and ceilings in the plant shall be of such construction as to be adequately cleanable and shall be kept clean and in good repair. Fixtures, ducts, and pipes shall not be so suspended over working areas that drip or condensate may contaminate foods, raw materials, or food-contact surfaces. Aisles or working spaces between equipment and walls shall be unobstructed and of sufficient width to permit employees to perform their duties without contamination of food or food-contact surfaces with clothing or personal contact.

(2) Provide separation by partition, location, or other effective means for those operations which may cause contamination of food products with undesirable microorganisms, chemicals, filth, or other extraneous material.

(3) Provide adequate lighting to hand-washing areas, dressing and locker rooms, and toilet rooms and to all areas where food or food ingredients are examined, processed, or stored and where equipment and utensils are cleaned. Light bulbs, fixtures, skylights, or other glass suspended over exposed food in any step of preparation shall be of the safety type or otherwise protected to prevent food contamination in case of breakage.

(4) Provide adequate ventilation or control equipment to minimize odors and noxious fumes or vapors (including steam) in areas where they may contaminate food. Such ventilation or control equipment shall not create conditions that may contribute to food contamination by airborne contaminants.

(5) Provide, where necessary, effective screening or other protection against birds, animals, and vermin (including but not limited to, insects and rodents).

§128.5 Sanitary facilities and controls.

Each plant shall be equipped with adequate sanitary facilities and accommodations including, but not limited to, the following:

(a) *Water supply.* . . .

(b) *Sewage disposal.* . . .

(c) *Plumbing.* . . .

(d) *Toilet facilities.* . . .

(e) *Hand-washing facilities.* . . .

(f) *Rubbish and offal disposal.* . . .

General Maintenance

In order to operate a sanitary food processing establishment, it is essential that the plant, equipment, and any associated physical facilities be maintained in good repair. A ramshackle, dilapidated building cannot be properly cleaned, and certainly cannot be kept sanitary.

Materials used in construction of the plant must be durable and cleanable. Fixtures, pipes, ducts, and so on, must be located and designed of such materials and in such a way that they do not collect dust or condensed moisture to contaminate food products in the same environment, and do not in and of themselves interfere with cleaning and housekeeping. Ventilation should be adequate, and if possible a positive air pressure should be maintained in any area where food is exposed to the air. This will keep airborne contaminants moving away from the food and prevent them from becoming a part of the food. In those areas where food is being handled, lighting should be adequate (as defined by GMPs, adequate means that which is needed to accomplish the intended purpose in keeping with good public health practice). The GMPs go into considerable detail as to protection of exposed foods from any possible contamination by broken glass as could occur from improperly designed lighting.

As has been mentioned in regard to many types of equipment, it should be designed and constructed of material that is durable, noncorrosive, non-absorbent, and easily cleanable. It should not have sharp or square corners, should have as few seams as can be accomplished, and certainly should not have cracks, crevices, or dead-end pockets. Equipment must be placed so that it is most easily used and cleaned, and if possible, so that it can be easily disassembled and assembled for cleaning and sterilization. As equipment in a plant ages, management should be considering new designs and placements to keep the plant in top sanitary conditions with modern equipment.

Animal and Vermin Control

The GMP regulations include:

(b) Animal and vermin control. No animals or birds, other than those essential as raw material, shall be allowed in any area of a food plant. Effective measures shall be taken to exclude pests from the processing areas and to protect against the contamination of foods in or on the premises by animals, birds, and vermin (including, but not limited to rodents and insects). The use of insecticides or rodenticides is permitted only under such precautions and restrictions as will prevent the contamination of food or packaging materials with illegal residues.

Insects, birds, or any animal pest around food processing or handling areas will constitute a source of contamination for the food. The presence of such pests is responsible for the filing of numerous complaints against food establishments, including food processing plants, each year. The complaints usually result when some insect or animal filth or parts have been discovered in a food product. Because such events usually get publicity, the general public has been made aware of the occurrence and the danger of such presence, and this results in increases in numbers of such complaints.

A major factor in control of any pest is the application of ordinary sanitation practices. These will remove food, water, and shelter for the pests, and will therefore keep them from moving into the plant, and starting to breed in this location. Pest control becomes a major occupation in any concentration of human population, because the pests thrive best in environments also inhabited by humans. An effective pest control measure is the correct screening of a plant, which will control insects, rodents, and birds. Filling holes, cracks, and crevices in a building will help to eliminate both rodents and insects.

Frequent inspection for the presence of pests is essential. Although the pest may not always be visible to an untrained observer, experience and practice will allow a sanitarian to recognize signs of infestation. When rodents are present, there will always be tell-tale presence of excreta pellets, and by the use of ultraviolet light, urine stains can be detected. After a period of infestation, body stains around walls, beams, and other places where nests may be located, or the presence of gnawed wood, paper, or cartons will indicate that the rodents are located within the environment. Some insects are more obvious, and can be observed in normal operational activities. Others, such as roaches, may be harder to find, but in any environment where these pests are located, there will be some sign of their presence. Darkened storage areas are generally the places where insects can be observed or heard.

When there is an indication that pests of any kind are present in the plant, the first step in control is to eliminate any possible breeding area outside the plant. After this has been done, screening and rat-proofing of the plant buildings should be re-inspected, and repaired where needed to prevent entrance of the pests into the plant. In addition to these measures, it is essential to take other measures as needed either continuously or on occasion to control pests. For flying insects, electrocutors can be a permanent control measure in certain plant locations. Pesticides with residual action can be painted on some surfaces, or sprayed in some locations, which will assist in control efforts. A thoroughly trained, knowledgeable individual must be in charge of this operation and any other controls within the plant, because there must be no residues in locations that may come in contact with the food products being processed. Rodent poisons may be necessary in some instances, and when this is the case, a licensed pest control operator should direct the control operations. In these cases, the anticoagulant rodenticides (i.e., warfarin, pival, etc.) can often be used with no difficulty. Infestation in cold storage areas can usually be eliminated by fumigation with harmless gases such as CO_2. Pests can generally be controlled with no great problems if the plant personnel are concerned about the control. Effective control includes consistent efforts at (a) environmental sanitation to remove sources of food and shelter, (b) screening and rat-proofing, and (c) extermination programs.

Cleaning and Sanitation

The GMPs as promulgated by the FDA do not specify a certain time interval for cleaning and/or sanitation of equipment and utensils, rather such statements as "all utensils and product-contact surfaces of equipment shall be cleaned as frequently as necessary to prevent contamination of food and food products" are included. Another statement is that "Single-service articles (such as utensils intended for one-time use, paper cups, paper towels, etc.) should be stored in appropriate containers and handled, dispensed, used, and disposed of in a manner that prevents contamination of food or food-contact surfaces."

It is specified that all equipment used in a food processing plant be cleaned and sanitized prior to use and following any interruption in use that could result in contamination. Unless such equipment is properly cleaned and sanitized, the product produced in a plant cannot be deemed a sanitary product produced under sanitary conditions. Trash around equipment is an invitation to leave soil around such equipment without frequent cleaning, which in turn is an invitation to insects and rodents to feed on the materials around, and will permit the multiplication of microorganisms in the area of food preparation and processing resulting in contaminated products.

Specific procedures for cleaning and sanitizing each piece of equipment must be written and available to the equipment operator at all times. When these procedures are designed, several factors must be considered, for example, the type of surface, the potential for food contact, the type of soil likely to be present on the surface, and whether the equipment is to be cleaned in place or is to be taken down and cleaned manually. Regardless of the procedure, it is essential that the personnel understand the use of the equipment, the necessity for keeping it clean, and the need to be cooperative with others in the handling, cleaning, and use of the equipment.

For example, if the users of equipment are cautious in maintaining the proper temperature, and avoiding burning of foods, cleaning is made much easier. Once use of the equipment has ended, proper and prompt rinsing can make cleaning easier because it avoids dried residues of foods. If the equipment has been properly maintained so that seals and gaskets have not allowed spillage, and the users work carefully in handling food products, the working area generally remains cleaner and the amount of strenuous cleaning required is greatly reduced.

It is necessary in continuous production operations that predetermined schedules using adequate cleaning and sanitizing procedures be in place and that the routine be faithfully followed. It is essential that the cleaning and sanitizing agents used be safe for the manner of use, and that the supervisors stay abreast of developments in the field so that the most efficient and effective agents are available for use. There are frequent and numerous improvements in all types of chemical agents, including those needed for cleaning and sanitizing in a food processing plant. Providing new and improved cleaning methods and agents to plant personnel will improve the performance of workers by improving morale as well as maintaining a safe product and environment. In July 1986, a bill was filed in the U.S. Congress to require the EPA to reinstate its product testing program in the area of sterilants, disinfectants, and antiseptics. The bill sought to amend the Federal Insecticide, Fungicide, and Rodenticide Act so that the EPA must "monitor and enforce efficacy standards for antimicrobial . . . agents used to control pest microorganisms which pose a threat to human health. . . ." It has been pointed out that the fact that the EPA registers disinfectants leads the public to believe that such materials are effective, when in fact some of the products as they are sold to the public are contaminated with microorganisms that are dangerous to human health.

When portable equipment and utensils are utilized in a plant, it is essential that a satisfactory storage area and schedule be arranged for such equipment when it is not in use. These items, after cleaning and sanitizing, must be stored so that product-contact surfaces are not splashed, contaminated, nor subject to settling dust. Although re-sanitization may be necessary before the next use, it should not be necessary to completely re-clean such equipment after it has been in temporary storage for only a short time.

Water Supply

Most ordinances require that the water used in any food processing plant must be from an approved source. In most cases, this means that the water supply must be, at the least, from the municipal supply which has been adequately treated, and is periodically tested. In food processing plants, the water may be used to add to the food; to clean raw foods; to clean equipment and utensils; to cool various articles, including food products; or as a means of conveyance for articles. Therefore, its quality must be beyond question at all times.

In food processing of any type, the amount of water required is generally extremely large. Much of this water may be returned to surface waters, and when this is so, it generally will carry large quantities of waste materials from the processed foods. Because so much water is needed, it is not uncommon in the food processing industry that plants will provide their own water supply. This is perfectly satisfactory, provided that the proper health authorities have approved the source, and do periodic testing of the water to assure that it is suitable for the intended use. One of the tests required for many food processors determines whether the mineral content of the water supply is satisfactory for inclusion in the food product to be processed or manufactured. Water hardness is also a factor that should be considered in food processing activities for several reasons, including those previously alluded to in chapter 6.

Waste Removal

Partly because of the effects on many surface water sources, and partly because of other potential undesirable effects, it is becoming increasingly important that food processors greatly reduce the amount of waste disposed of in surface waters, and that they control the type of waste that may be disposed of in this manner. Many food wastes are potentially good culture media for bacterial populations, including those that may be present in surface waters, and therefore the addition of these wastes to the surface supply may simply permit the culture of huge populations of microorganisms that may or may not be harmful or beneficial to the eventual user of the water.

Previously, wastes generated in food processing could be dealt with locally, and there was no great harm done to the environment where the waste was produced. For example, a dairy plant processing between 2,000 and 3,000 quarts of milk per day can remove waste and feed it to pigs locally, solving two problems. However, a plant producing tons of wastes each year cannot solve the waste disposal problem so easily. The tons of wastes produced by large dairy plants each year should be further processed and uses found for some of the wastes such as in fermentation, production of carbohydrates, production of ethanol, and production of useful proteins. If such uses are found, then wastes become a useful raw material, and can be used for production of needed products.

Generally, wastes can be handled more easily and further processed more readily if most of the water has been removed. Of course in doing this, some waste escapes as liquid, but the remainder is a useful product, and does not have to be disposed of in the environment. In the citrus juice industry, more than ninety-nine percent of the raw material is utilized as some useful product, from juice to cattle feed.

Personnel

The most modern equipment, and the best drawn plans for a sanitary system will be useless unless the people involved are correctly trained, and unless they are interested enough to put the plans into effect properly. The sanitation program will be only as effective as the persons who carry it out. To do this correctly, then, the personnel must receive the proper training, and supervision. The sanitarian must not only have the knowledge required for his job, but must also be able to teach others so that other workers can help to carry out the program. The sanitarian should have facilities, schedules, and equipment to carry on regular training sessions for all workers concerned with the program, and to keep these workers updated with a frequency that matches changes in the industry.

CONTROL

The World Health Organization Expert Committee on Microbiological Aspects of Food Hygiene has put together a program termed Hazard Analysis and Critical Control Point (HACCP). This is a program to be studied and elaborated for use in both developed and developing countries for control of health hazards in food establishments. The committee set out to review the literature relating to HACCP, collect information on the practical use of HACCP, and assess the practicality of its use in developed countries as well as developing a plan for its potential use in developing countries. The committee reported in 1980 with a plan that has been widely distributed and used since that time. Among the statements in the report were that mishandling of foods causing outbreak of infectious diseases was much more common in the home than in food processing plants, or foodservice establishments. The occurrence of such mishandling was least common in food processing establishments, but it was pointed out that the potential for affecting large numbers of individuals when food is contaminated is much greater in a food processing establishment than in a foodservice establishment or in a home.

The control of microbiological hazards in foods has been accomplished, and continues to be effected by a combination of education and training, inspection of facilities and operations, and microbiological testing. Programs that utilize all three of these factors are generally successful in avoiding the occurrence of foodborne disease outbreaks. In these factors, the process of education should be aimed at teaching the causes, consequences, and avoidance of microbial contamination, as well as teaching personal hygiene, community sanitation, and food sanitation. The last three items should have been learned during early phases of life and in primary education experiences, but should definitely be reinforced, and reiterated in the training phases utilized by food processing operations. Obviously, different, broader and more in-depth training is required for supervisory personnel than for some personnel in the more routine jobs.

It is essential that facilities, equipment, and routine practices be inspected by supervisory personnel and by regulatory authorities in some instances. If such sanitation inspections are performed thoroughly and as part of a regular and reasonable routine, then it is generally the case that the desired level of cleanliness and sanitation is maintained in all operations. The inspector must require good hygienic routine practices in all operations, and can usually provide to the personnel specific documents requiring certain operations from GMP guidelines, or other codes that have been developed by experts in the field. Because these guidelines generally are deliberately written using rather vague terms, for example, *satisfactory, acceptable,* it is sometimes necessary at the discretion of the inspector to provide more specific detail, or at least to determine acceptable practices. It is therefore necessary that the inspector be thoroughly versed in the important versus unimportant points in operations from the standpoint of hazards reduced. Because the importance of these inspections is so great, it may be well for the plant to employ a committee or a team for regular inspections so that if one item is missed by one person, it may be picked up by another. Inspections are generally made according to specified items and practices, and the report may very well be made on a standardized form made up accordingly. If this is the case, there should be some margin for variation or comments by the inspecting team so that they are not bound by circumstances that do not fit their operation. If deficient items are found, the specifics of the deficiency *must* be stated so that corrective action can be detected on the next inspection.

The size of the operation, and the number of different operations in the process will influence the forms used, and the nature and formality of inspections. It is necessary, however, that frequent informal inspections be held between the formal ones, in order that defects in the operation may be detected before real harm is done, and before an outbreak of infectious disease has been initiated by the defect.

To substantiate the validity of the visual inspection, as well as to give specific information, products, containers, and equipment should also be examined by laboratory microbiological testing. If these product tests are done as a part of the processing routine, as well as in unscheduled spot checks, then it is most likely that any defect or deficiency in the operation will be detected before it has caused any serious damage to the product. These tests help to determine whether personnel have adhered to GMP guidelines, and frequently will detect the exact place in the process where there has been a failure or contamination of the product.

It is essential that personnel and particularly supervisory personnel be completely versant with the types, characteristics, and sources of different microorganisms that may be encountered in testing programs. It must be understood that simple total counts of microbes present may not in every case bear a direct relationship to spoilage potential or to the transmission of infectious disease by the product. It is essential that the interpreter of the data know the differences between indicator presence, pathogen presence, thermophile presence, psychrophile presence, and mesophile presence. For example, in canning, or in some smoking operations, it is important that the personnel involved be knowledgeable about the significance of the presence of numbers of spore formers among the total bacterial populations. The purposes of the microbial testing is to detect defects in the product in terms of potential for spoilage, potential for transmission of infectious disease, and potential for increasing or reducing shelf life of the product. If the microbiological testing is done along the line of production, then the condition of the product at any one stage can be detected, and any trouble spots that occur can be pinpointed. When these are determined, then it is not necessary that the entire production process be reexamined and reestablished in order to eliminate the defect.

To establish a HACCP system for a given food processing operation, it is necessary that there be (a) an assessment of hazards associated with the processing or manufacturing program; (b) determination of critical control points required to control those hazards that were identified; and (c) a system to monitor the critical control points established. These three steps will be specifically different for food processing establishments, and must be determined internally for each operation, although each step does have some common features that can be followed to establish the system.

In the assessment of hazards, it is essential that each procedure associated with the production, distribution, or use of raw materials or foods be critically analyzed to (a) identify all potentially hazardous raw materials or foods; (b) identify the source and the specific point of contamination by looking at each step in the process; and (c) identify the potential for any microbial population to survive/multiply during processing or in storage. For this purpose, (a) includes any raw or processed food that may contain any toxic substance, pathogen, or large numbers of microorganisms, or a food that may very effectively support the growth of such microorganisms. The factors included in (b) and (c) are self-explanatory. In food processing, all factors and changes in raw materials used (state, source, condition), processing formula, packaging system or materials, distribution system, storage location or conditions, and product nature must be included in the hazard analysis. Because the microbiological hazards vary from one product to another, and from one method of handling or merchandising to another, it is important that such analyses be conducted for each food processing operation, even when expanding a single operation to a new location.

Microbiological contamination of raw materials must be evaluated separately in every instance. In some cases, as in the blending of dry products, processing cannot be relied upon to remove or limit contamination that may be in the raw materials. On the other hand, any product manufactured from animal materials must be assumed to be contaminated with pathogenic microorganisms, and the process used must eliminate such contaminants. Processes used in food plants may include heat treatment, acidification, fermentation, and salting that will kill or inhibit the growth of microorganisms present. However, some handling steps following these processes may actually recontaminate the product (i.e., cooling, boning, slicing). These potential trouble spots are further factors in the needs for product testing at different steps and stages of processing. To assure the effectiveness of the hazard analysis program, the plant personnel should be alert to recognize any and all practices that might potentiate cross-contaminations, air currents blowing on finished products, contamination of product by cooling water, and so on.

Food plant sanitation programs are, of necessity, never-ending. It is essential to the safe operation of a food processing plant that the plant operators, from top management, to the lowest paid worker, be constantly *aware* and alert to the correct operation of a properly designed sanitation program. In this operation, the sanitarian is simply the one employee who is *solely* involved in the plant sanitation program. Unless other employees are also involved, and on a daily and constant basis, the sanitation program has little chance for success.

BIBLIOGRAPHY

FOX, J. L.   1986.   Renewed concern to monitor and test disinfectants. *ASM News* 52:534–536.

INTERNATIONAL COMMITTEE ON MICROBIOLOGICAL SAFETY OF FOOD.   1980.   Hazard analysis and critical control point: application to food safety and quality assurance in developed and developing countries. Report to the WHO Expert Committee on Microbiological Aspects of Food Hygiene. Geneva, Switzerland.

LECOS, C. W.   1986.   The growing use of irradiation to preserve food. *FDA Consumer* (October: 12–15).

RIEMANN, H., and BRYAN, F. L., eds.   1979.   *Food-borne infections and intoxications.* 2nd ed. New York: Academic Press.

TROLLER, J. A.   1983.   *Sanitation in food processing.* New York: Academic Press.

UNITED NATIONS ENVIRONMENT PROGRAM.   1985.   The food processing industry and the environment. *Industry and Environment* 8:1–3.

UNITED STATES DEPARTMENT OF HEALTH, EDUCATION, AND WELFARE, FDA.   1974.   *Human foods: Current good manufacturing practice (sanitation) in manufacture, processing, packing, or holding.* Part 128. 21 CFR. Washington, DC: U.S. Government Printing Office.

11

Foodservice Sanitation

In 1935 the U.S. Public Health Service published tentative guidelines for ordinances and codes to regulate foodservice establishments. These guidelines were intended to assist state and local governments and health departments in promulgating specific ordinances to govern establishments in their jurisdiction. Those guidelines have been used to a great extent by local authorities as a model for ordinances. There have been relatively few changes, and the guidelines have not been updated since 1976, although there have been great changes in food technology, eating habits, mode of living, and economic conditions throughout the world. Particularly in the United States, foodservice establishments have grown in number, variety, size, and quality, and have varied in stability in recent years, but in general, the model for foodservice ordinances has continued to work very well. The success of federal, state, and local law enforcement in these regulations and ordinances has contributed greatly to the success of foodservice establishments by regulating production, supply, processing, preparation, and service of safe foods of good quality.

PHYSICAL FACILITIES

A major portion of the *Food Service Sanitation Manual* (1976) deals with the construction and maintenance of physical facilities used in the process of foodservice. Items dealt with in this aspect include the construction materials, maintenance, and use of floors, walls, ceilings, lighting, ventilation, dressing rooms, locker rooms, storage (particularly storage of toxic materials), living areas, laundry facilities, and animals on the premises. A major consideration in the recommendations is the ease of repair, cleaning, and sanitizing within the establishment. Types of satisfactory materials are specified, and in the case of floors, a major concern is the use of carpeting, and the use of anti-slip floor coverings where needed for safety. Carpeting is prohibited in areas to be used for food preparation, ware-washing, storage, and toilets. Also prohibited within the establishment are floors covered with sawdust, wood shavings, peanut hulls, or any similar materials that would be a major source of dust in the environment. The need for floor drains in certain areas is recognized, and specifics for their installation are recommended. The prohibition of utility service lines and pipes on the floor is recommended. The reasoning behind all recommendations is explained in the manual dealing with the specific items. In this way, the recommendations may be easily understood by persons not having the direct experience of working in such an environment, that is, most of those persons who will be involved in the writing of laws or ordinances that will govern the operations of foodservice establishments.

How closely these recommendations are followed may be observed in the following section extracted from the *Rules and Regulations for Food Service from the State of Georgia;* Section 290-5-14-.18, Other Facilities and Operations Amended:

(1) Floors, Walls, and Ceilings:

(a) Floors

1. All floors shall be kept clean and in good repair. Sawdust or wood shavings shall not be used on the floors.

2. The floors of all food preparation, food storage, and utensil-washing rooms and areas, and walk-in refrigerators, dressing or locker rooms, and toilet rooms shall be constructed of smooth, durable, nonabsorbent, and easily cleanable materials such as concrete, terrazzo, ceramic tile, durable grades of linoleum, or tight wood impregnated with plastic: provided, that in areas subject to spilling or dripping of grease or fatty substances, such floor coverings shall be of grease-resistant material; and provided further, that floors of nonrefrigerated, dry-food-storage areas need not be nonabsorbent.

3. Floor drains shall be provided in floors which are water-flushed for cleaning or which receive discharges of water or other fluid waste from equipment. Such floors shall be graded to drain.

4. Carpeting, fire resistant and water repellant, may be used on the floors of interior dining areas. Such carpeting shall be in good repair and kept clean.

5. The walking and driving surfaces of all exterior areas where food is served, such as drive-in restaurants, side-walk cafes, patio service, chuck-wagon service, and barbecues, shall be kept clean and free of debris, and shall be properly drained so that water will not accumulate. Such areas shall be surfaced with concrete or asphalt, or with gravel or similar material effectively treated to facilitate maintenance and to minimize dust.

6. Mats or duckboards, if used, shall be so constructed as to facilitate being cleaned, and shall be kept clean. They shall be of such design and size as to permit easy removal for cleaning.

7. All concrete, terrazzo, or ceramic tile floors, hereinafter installed in food

preparation, food storage, and utensil-washing rooms and areas, and in walk-in refrigerators, dressing or locker rooms, and toilet rooms, shall provide a coved juncture between the floor and wall. In all cases, the juncture between the floor and wall shall be closed.

The recommendations concerning the structure of walls and ceilings are made so that they are properly constructed, easily cleaned, and can be maintained in good repair so that they are nonabsorbent and will not create unsanitary conditions. Some of the specifics that will be found in laws dealing with these structures are observed in the following section from the State of Texas Rules for Food Service Sanitation:

(b) Walls and Ceilings.

(1) Maintenance. Walls and ceilings, including doors, windows, skylights, and similar closures, shall be maintained in good repair.

(2) Construction. The walls, including nonsupporting partitions, wall coverings, and ceilings of walk-in refrigerating units, food preparation areas, food storage areas, equipment-washing and utensil-washing areas, toilet rooms and vestibules shall be light-colored, smooth, nonabsorbent, and easily cleanable. Concrete or pumice blocks used for interior wall construction in these locations shall be finished and sealed to provide an easily cleanable surface.

(3) Exposed Construction. Studs, joists, and rafters shall not be exposed in those areas listed in paragraph (2) of subsection (b) of this rule. If exposed in other rooms or areas, they shall be finished to provide an easily cleanable surface.

(4) Utility Line Installation. Exposed utility service lines and pipes shall be installed in a way that does not obstruct or prevent cleaning of the walls and ceilings. Utility service lines and pipes shall not be unnecessarily exposed on walls or ceilings in those areas listed in paragraph (2) of subsection (b) of this rule.

(5) Attachments. Light fixtures, vent covers, wall-mounted fans, decorative materials, and similar equipment attached to walls and ceilings shall be easily cleanable and shall be maintained in good repair.

(6) Covering Material Installation. Wall and ceiling covering materials shall be attached and sealed so as to be easily cleanable.

Although the foodservice operation must be kept clean to minimize attraction of pests, and to prevent contamination of foods, the cleaning of such a business must be done in a way such that potential for contamination is not increased. The manual recommends the following:

7-301 General.

Cleaning of floors and walls, except emergency cleaning of floors, shall be done during periods when the least amount of food is exposed, such as after closing or between meals. Floors, mats, duckboards, walls, ceilings, and attached equipment and decorative materials shall be kept clean. Only dustless methods of cleaning floors and walls shall be used, such as vacuum cleaning, wet cleaning, or the use of dust-arresting compounds with brooms.

7-302 Utility Facility.

In a new or extensively remodeled establishment at least one utility sink or curbed cleaning facility with a floor drain shall be provided and used for the cleaning of mops or similar wet floor cleaning tools and for the disposal of mopwater or similar liquid wastes. The use of lavatories, utensil-washing or equipment-washing, or food preparation sinks for this purpose is prohibited.

Adequate lighting in all parts and corners of the foodservice operation will help to make dirt conspicuous to the patrons of the establishment, and is also necessary for proper cleaning of all areas. If lighting is inadequate, it may help to establish a pattern of operation that will lead to a general lack of cleanliness, and a lack of attention to detail in food preparation that may result in accidental erroneous mixing of ingredients, or even to serving obviously contaminated or spoiled dishes to customers. It is obvious that unprotected light fixtures above food preparation or service counters are a potential source for contamination by glass fragments if breakage occurs. It is therefore likely that many local and state rules will contain sections similar to the following from the *Food Ordinance of the City of Houston, Texas.*

LIGHTING;

General. At least 50 foot-candles of light shall be provided to all working surfaces and at least 30 foot-candles of light shall be provided to all other surfaces and equipment in food-preparation, utensil-washing, and hand-washing areas, and in toilet rooms. At least 20 foot-candles of light at a distance of 30 inches from the floor shall be provided in all other areas, except that this requirement applies to dining areas only during cleaning operations.

Protective shielding.

(a) Shielding to protect against broken glass falling onto food shall be provided for all artificial lighting fixtures located over, by, or within food storage, preparation, service, and display facilities, and facilities where utensils and equipment are cleaned and stored.

(b) Infra-red or other heat lamps shall be protected against breakage by a shield surrounding and extending beyond the bulb, leaving only the face of the bulb exposed.

The above exceeds the recommendation for lighting in other areas in the *Food Service Sanitation Manual of 1976,* which recommends at least ten foot-candles of light in walk-in refrigerating units, dry food storage areas, and in all other areas, including dining areas during cleaning.

The manual recommends adequate "ventilation to keep rooms free of excessive heat, steam, condensation, vapors, obnoxious odors, smoke and fumes. Ventilation systems shall be installed and operated according to law and, when vented to the outside, shall not create an unsightly, harmful or unlawful discharge." The reasons for these recommendations are to prevent condensate from dropping into food or onto food preparation areas, and to sufficiently remove moisture so that the growth of molds and bacteria will not be encouraged.

The wearing of street clothing in work areas of foodservice establishments creates conditions whereby food and food preparation surfaces can be contaminated by microorganisms from outside the building. Therefore, the manual recommends that dressing rooms and locker rooms be provided for employees, and prohibits these spaces from being used for food preparation, storage, or service. The manual also recommends that poisonous or toxic materials be permitted in foodservice businesses only for the purpose of maintaining the business in a clean and sanitary condition, and for controlling pests. When such materials are present, they should be properly and clearly labeled, and must be stored in a facility completely separate for each type, and completely separate from all other stored supplies in the foodservice building. The use of these toxic materials is restricted to use in such a way that neither food nor food contact surfaces are contaminated or left with harmful residues. It is recommended that personal medications be kept stored in a separate area, probably locker room areas, with the personal belongings of employees, and that these never be stored in areas used for food preparation, storage, or service. Although first-aid supplies may be kept in a foodservice establishment, they should always be kept in an area, and in such a way that prevents contamination of food or supplies.

The recommendations would prevent use of foodservice establishments for living quarters for humans or animals. The use of areas for laundry and for linen storage, or cleaning supply storage should be limited, and not be used for handling of food in any manner. Excess, or unnecessary traffic of persons not engaged in food preparation or service for the establishment should be severely limited, or prohibited, and the only animals allowed in foodservice establishments should be seeing eye dogs.

In following the recommendations of the *Food Service Sanitation Manual,* ordinances should specify that accumulation of dust, dirt, and filth, including food scraps on floors, walls, ceilings, or fixtures is strictly forbidden and that such surfaces be easily cleanable, and that such cleaning be done at frequent intervals to maintain the aesthetic as well as the sanitary condition of the establishment. Also as mentioned in an earlier paragraph, the cleaning must be done in a manner that will not create dust and contamination, and should be done when food is not being prepared, and when customers are not eating.

Following superficial cleaning such as vacuuming, it is important that wherever possible, floors be washed and disinfected at regular intervals. The act of sweeping by any method is totally inefficient as far as sanitization is concerned, and some food particles and much microbial contamination will remain. To avoid attracting insects and inviting microbial multiplication, disinfection must be included in the sanitation routine, and this routine must include corners, entrances, and crevices. The use of insecticide paints at intervals will assist in preventing infestation with these pests. Carpets must be vacuumed because other sweeping methods are totally ineffective. When carpeting is present in service areas, it is most imperative that cleaning be done on a frequent schedule because liquids and foods in the carpet nap will provide an excellent environment for microbial growth, and the holding qualities of the carpet will serve as bait for insect attraction. Improved types of carpet that are most easily cleaned and are nonabsorbent are much preferred for foodservice areas. These are not readily stained, and usually permit a fairly wide variety of cleaning chemicals to be used, some of which have a fair degree of disinfectant action. Frequent use of such chemicals should be a standard sanitation procedure in foodservice establishments. In some areas it has become fairly common to carpet walls, and to some extent such a practice would not create the same environment for contamination as would floors. However, in the lower walls around foodservice areas, splashing and contamination with foods and drinks would be fairly common, and where this is possible, it must be taken into account in the cleaning and sanitizing routine of the business. Otherwise, walls and ceilings would not require attention as frequently nor as intensely as would floors in foodservice establishments. The routines must be done often enough to prevent the accumulation of grease, smoke, and grime, which will spoil both the aesthetic and the sanitary environment.

WATER SUPPLY

The recommendations issued in the *Food Service Sanitation Manual, 1976* concerning water supply are rather brief, but specific. These state "Enough potable water for the needs of the food service establishment shall be provided from a source constructed and operated according to law." They go on to say "All potable water not provided directly by pipe to the food service establishment from the source shall be transported in a bulk water transport system and shall be delivered to a closed-water system. Both of these systems shall be constructed and operated according to law." And,

All bottled and packaged potable water shall be obtained from a source that complies with all laws and shall be handled and stored in a way that protects it from contamination. Bottled and packaged potable water shall be dispensed from the original container. Water under pressure at the required temperatures shall be provided to all fixtures and equipment that use water. Steam used in contact with food or food-contact surfaces shall be free from any materials or additives other than those specified in 21CFR 173.310.

The latter reference refers to the current code of Federal Regulations dealing with allowed additives for cleaning or sanitizing in areas that could contaminate food or food contact surfaces. Both the *Texas Department of Health Rules on Food Service Sanitation* and the *Houston Food Ordinance* copy these recommendations verbatim; and the Food Ordinance then adds the following sections dealing with water fountains and ice manufacture.

Water fountains. Water fountains shall be constructed and installed according to all applicable laws and maintained clean and in good repair.
Ice manufacture. Ice shall be made from water meeting the requirements of . . . the section . . . *Water Supply,* in an approved ice-making machine or plant which is located, installed, operated, and maintained so as to prevent contamination of the ice; or shall be obtained from a source approved by the health officer.

The manual does differentiate between stationary foodservice establishments and those offering mobile or temporary foodservice. For a mobile system:

A mobile food unit requiring a water system shall have a potable water system under pressure. The system shall be of sufficient capacity to furnish enough hot and cold water for food preparation, utensil cleaning and sanitizing, and handwashing, in accordance with the requirements of this ordinance. The water inlet shall be located so that it will not be contaminated by waste discharge, road dust, oil, or grease, and it shall be kept capped unless being filled. The water inlet shall be provided with a transition connection of a size or type that will prevent its use for any other service. All water distribution pipes or tubing shall be constructed and installed in accordance with the requirements of this ordinance.

For a temporary facility: "Enough potable water shall be available in the establishment for food preparation, for cleaning and sanitizing utensils and equipment, and for handwashing. A heating facility capable of producing enough hot water for these purposes shall be provided on the premises."

PLUMBING

The requirements for materials, installation codes, and numbers, kinds, and location of fixtures as well as safety features vary in different municipalities. Improper plumbing can result in cross-connections, back siphons, and leaks. These will result in contamination of the establishment, and most likely contamination of food contact surfaces, or even food supplies themselves. The general recommendations of the *Food Service Sanitation Manual* are as follows:

Plumbing shall be sized, installed, and maintained according to law. There shall be no cross-connections between the potable water supply and any nonpotable or questionable water supply nor any source of pollution through which the potable water supply might become contaminated.

A nonpotable water system is permitted only for purposes such as air-conditioning and fire protection and only if the system is installed according to law and the nonpotable water does not contact, directly or indirectly, food, potable water, equipment that contacts food, or utensils. The piping of any nonpotable water system shall be durably identified so that it is readily distinguishable from piping that carries potable water.

The potable water system shall be installed to preclude the possibility of backflow. Devices shall be installed to protect against backflow and back siphonage at all fixtures and equipment where an air gap at least twice the diameter of the water supply inlet is not provided between the water supply inlet and the fixture's flood level rim. A hose shall not be attached to a faucet unless a backflow prevention device is installed.

If used, grease traps shall be located to be easily accessible for cleaning.

If used, garbage grinders shall be installed and maintained according to law.

Except for properly trapped open sinks, there shall be no direct connection between the sewerage system and any drains originating from equipment in which food, portable equipment, or utensils are placed. When a dishwashing machine is located within 5 feet of a trapped floor drain, the dishwasher waste outlet may be connected directly on the inlet side of a properly vented floor drain trap if permitted by law.

The *State of Georgia Rules and Regulations for Food Service* follow the general theme of the preceding recommendations, if not the exact wordage.

1. All plumbing shall be sized, installed, and maintained in accordance with applicable state and local plumbing laws, ordinances, and regulations.

2. The potable water-supply piping shall not be directly connected with any nonpotable water-supply system whereby the nonpotable water can be drawn or discharged into the potable water-supply system. The piping of any nonpotable water system shall be adequately and durably identified, such as by distinctive yellow-colored paint, so that it is readily distinguished from piping which carries potable water; and such piping shall not be connected to equipment or have outlets in the food-preparation area.

3. The potable water system shall be installed in such a manner so as to preclude the possibility of backsiphonage.

UTENSILS AND EQUIPMENT

Design and Location

Regulation of utensils and equipment for foodservice by local and state health authorities usually consists of multiple rules dealing with the construction, maintenance, frequency and methods of cleaning, disinfection, storage, and handling of these tools. Specific requirements for utensils may typically include a statement that this equipment be constructed of nontoxic material, and in some ordinances, the use of cadmium or lead is specifically prohibited. In addition to the material used, almost invariably the rule includes a requirement that the utensils be constructed in such a manner that they may be easily cleaned and inspected in all parts. Such utensils should have linings of metal, ceramic, or porcelain that will not be broken down by heat and that will not release harmful chemicals or metals into foods. Seams, crevices, and joints that come into contact with foods should be minimal, and should be constructed so that they are easily cleaned. Utensils and equipment should be kept in good repair and should not be used when they are "chipped, cracked, rusted, corroded, badly worn, or in such condition that they cannot be easily rendered clean and sanitary" according to one municipal ordinance. The manual recommends that if solder is used in the construction of equipment or utensils, it should be composed of safe materials and should be corrosion resistant. For cutting blocks, boards, salad bowls, and bakers' tables, either hard maple or some other wood that is equally nonabsorbent should be used. Other woods may be used for single-service articles, such as chopsticks, stirrers, or ice cream spoons, but reuse of single-service articles is to be prohibited in all foodservice regulations. When plastics or rubber articles are used, the material should be resistant to scratching, scoring, decomposition, and so on, and articles should be of enough weight and thickness to permit cleaning and sanitizing by normal dishwashing procedures. If mollusk or crustacean shells are used as serving containers, they may be used once only, and then must be discarded.

The design and fabrication of equipment and utensils is described here in the recommendations of the *Food Service Sanitation Manual of 1976.*

All equipment and utensils, including plasticware, shall be designed and fabricated for durability under conditions of normal use and shall be resistant to denting, buckling, pitting, chipping, and crazing.

(a) Food-contact surfaces shall be easily cleanable, smooth, and free of breaks, open seams, cracks, chips, pits, and similar imperfections, and free of difficult-to-clean internal corners and crevices. Cast iron may be used as a food-contact surface only if the surface is heated, such as in grills, griddle tops, and skillets. Threads shall be designed to facilitate cleaning; ordinary "V" type threads are prohibited in food-contact surfaces, except that in equipment such as ice makers or hot oil cooking equipment and hot oil filtering systems, such threads shall be minimized.

(b) Equipment containing bearings and gears requiring unsafe lubricants shall be designed and constructed so that the lubricant cannot leak, drip, or be forced into food or onto food-contact surfaces. Only safe lubricants shall be used on equipment designed to receive lubrication of bearings and gears on or within food-contact surfaces.

(c) Tubing conveying beverages or beverage ingredients to dispensing heads may be in contact with stored ice: Provided, That such tubing is fabricated from safe materials, is grommeted at entry and exit points to preclude moisture (condensation) from entering the ice machine or the ice storage bin, and is kept clean. Drainage or drainage tubes from dispensing units shall not pass through the ice machine or the ice storage bin.

(d) Sinks and drain boards shall be self-draining.

If equipment and utensils do not meet the general requirements listed in the preceding, they are difficult to clean and sanitize, and permit accumulations of foods and soils that will support the growth of contaminating bacteria. The equipment used must be easily accessible and meet the following requirements in this respect; "Unless designed for in-place cleaning, food-contact surfaces shall be accessible for cleaning and inspection:

(a) Without being disassembled; or

(b) By disassembling without the use of tools; or

(c) By easy disassembling with the use of only simple tools such as a mallet, a screwdriver, or an open-end wrench kept available near the equipment."

Permanently installed equipment, or even some moveable equipment may be either table- or floor-mounted. Some requirements are equally important for both types, but some are specific for the equipment involved. If equipment is permanently installed, it must be constructed in such a way that those portions that contact foods can be removed or reached for cleaning and sanitizing. Included in equipment of this type would be slicers, can openers, mixers, grinders, and bottle or can sealers. Thorough cleaning and disinfection of such equipment is necessary to prevent the spread of contamination throughout a food preparation area. For example, passage of one food contaminated with small numbers of *Salmonella* organisms from a carrier's hands will leave an inoculum of *Salmonella* on a food grinder, on the cutting blades, housing, switches, and other places. The ground food may be cooked to kill those organisms present before it is served; however, those contaminating cells remaining on the equipment are incubated at room temperature and continue to multiply. A cursory rinsing of the equipment will leave considerable numbers of bacterial cells in more protected areas to contaminate succeeding foods that are ground in the same machine, and will likely leave external contamination to be transferred to the hands of workers and thus will be spread to other food preparation areas. Such equipment should be thoroughly washed and disinfected at frequent intervals. Many municipal ordinances specify daily cleaning for utensils and equipment, which will suffice for most situations. If, however, there is increased likelihood of contamination, as in the case of contact with raw foods that may carry microorganisms, this interval may be shortened to as often as after each use.

Many health codes specify standards that must be met in the design and construction of specialized equipment such as steam tables, automatic dishwashers, refrigeration equipment, food dispensers, and foodservice equipment, which generally are models of floor-mounted equipment. Standards that are set for such items are those that ensure efficient cleaning, effective temperature control, and effectiveness of operation in the purpose for which the equipment was designed. In localities where these standards are set, rigid compliance with the regulations provides adequate protection to maintain sanitation. Where standards are not set, close inspection to determine cleanability and check for reliability of operation is necessary. As in the case of utensils, minimum seams, crevices, corners, and maximum availability for cleaning must be considered.

In addition to location and accessibility of the equipment itself, consideration must be given to aisles and spaces between equipment units as recommended by the manual. "Aisles and working spaces between units of equipment and walls shall be unobstructed and of sufficient width to permit employees to perform their duties readily without contamination of food or food-contact surfaces by clothing or personal contact. All easily moveable storage equipment such as pallets, racks, and dollies shall be positioned to provide accessibility to working areas."

It is not uncommon that cleaning, sanitization, and storage of equipment and utensils be considered together; however, some items must be considered separately. The frequency of cleaning and sanitization varies depending upon the utensils or equipment involved, and upon the specific use of those items.

(a) Tableware shall be washed, rinsed, and sanitized after each use.

(b) To prevent cross-contamination, kitchenware and food-contact surfaces of equipment shall be washed, rinsed, and sanitized after each use and following any interruption of operations during which time contamination may have occurred.

(c) Where equipment and utensils are used for the preparation of potentially hazardous foods on a continuous or production-line basis, utensils and the food-contact surfaces of equipment shall be washed, rinsed, and sanitized at intervals throughout the day on a schedule based on food temperature, type of food, and amount of food particle accumulation.

(d) The food-contact surfaces of grills, griddles, and similar cooking devices and the cavities and door seals of microwave ovens shall be cleaned at least once a day; except that this shall not apply to hot oil cooking equipment and hot oil filtering systems. The food-contact surfaces of all cooking equipment shall be kept free of encrusted grease deposits and other accumulated soil.

(e) Non-food-contact surfaces of equipment shall be cleaned as often as is necessary to keep the equipment free of accumulation of dust, dirt, food particles, and other debris.

The *Dallas City Ordinance* includes the following section:

CLEANING, SANITIZATION AND STORAGE OF EQUIPMENT AND UTENSILS

Section. 17-5.1. *Cleaning Frequency.*

(a) A food products establishment shall wash, rinse, and sanitize:

(1) tableware after each use;

(2) food-contact surfaces of equipment:

(A) before and after each product change;

(B) after processing raw and before processing a cooked or finished product; and

(C) following an interruption of operations during which contamination could occur;

(3) food-contact surfaces of grills, griddles, and similar cooking devices once each day, except hot oil cooking equipment and hot oil filtering systems;

(4) the cavities and door seals of microwave ovens once each day;

(5) food-contact surfaces of cooking equipment as often as necessary to

keep the equipment free of encrusted grease deposits and other accumulated soil.

(b) A food products establishment shall wash and rinse non-food-contact surfaces as often as necessary to keep the equipment free of dust, dirt, food particles, and other debris.

Section. 17-5.2. *Approved Chemical Sanitizing Agents.*

The director shall maintain a list of approved chemical sanitizing agents and additives that he has determined may be safely used on food-contact equipment and utensils along with proper temperatures and concentrations for use, and shall distribute and make available the list to permittees under this chapter.

Section. 17-5.3. *Wiping Cloths and Sponges.*

(a) A food products establishment that uses cloths for wiping food spills or tableware, such as plates or bowls being served to the consumer, shall maintain the cloths in a clean, dry condition and use them for no other purpose.

(b) A food products establishment that uses moist cloths or sponges for wiping food spills on kitchenware and food-contact surfaces of equipment shall:

(1) clean and rinse the cloths or sponges frequently in a sanitizing solution permitted in Section 17-5.2;

(2) use the cloths or sponges for no other purpose; and

(3) store the cloths or sponges in the sanitizing solution between uses.

(c) A food products establishment that uses moist cloths or sponges for cleaning non-food-contact surfaces of equipment, such as counters, dining table tops, and shelves, shall:

(1) clean and rinse the cloths or sponges frequently in a sanitizing solution permitted in Section 17-5.2;

(2) use the cloths or sponges for no other purpose; and

(3) store the cloths or sponges in the sanitizing solution between uses.

Specific methods, chemicals, and temperatures are required for manual cleaning, and for mechanical cleaning and sanitizing, only specifically approved chemical agents are to be used (see Tables 11.1

Table 11.1.   Cleaning Agents

1. Water—Clean/Potable—Solvent for soil and chemicals
2. Strong alkalis—Detergents for fats and proteins—Precipitate hardness— Corrosive
3. Mild alkalis—Detergents, buffers—Mildly corrosive
4. Inorganic acids—Remove precipitates on surfaces—Very corrosive and skin irritant
5. Organic acids—Remove inorganic precipitates—Mildly corrosive
6. Wetting agents—Anionic/nonionic/cationic—Detergents—Emulsifiers—Not all compatible with each other
7. Sequestering Agents—Phosphates—Complex Metals—Cannot be heated or exposed to acids
8. Abrasives—Use with detergents—Scratch and etch surfaces
9. Enzymes—Digest proteins—Cannot be heated
10. Others—Chlorinated Compounds—Amphoterics—Little used

Table 11.2.   Representative Sanitizing Agent Groups

Agent Group	Properties
Hypochlorites	Leave no dangerous residue
Iodides, iodines	Leave no dangerous residue
Sulfonic acid preparations	Leave no dangerous residue
Iodine polyalkylene glycol	Cannot be used as final rinse
Others	No final rinse required
Ethylene propylene polyalkylene glycol	
Polyoxyethylene, polyoxypropylene	
Dodecylbenzenesulfonic acid	
Butoxy polyalkylene glycol	
N-Alkyl benzyldimethylammonium chloride	
Trichloromelamine	
Sulfonated oleic acid	
Lithium hypochlorite	

The compounds listed are representative chemicals. Specific chemical mixtures and final rinses are specified for certain usages in local rules and regulations.

and 11.2). Cleaning-in-place procedures and chemicals are specified for most equipment types that can be cleaned in this manner. The mechanical equipment for sanitizing is required to maintain operation at certain minimum temperatures and conditions. In any case, if sanitizing is to be done effectively, washing away of food masses is necessary before final treatment because microorganisms encased in masses of food or organic matter will not be reached by chemicals, and frequently will not be subjected to heat sufficient for killing. Procedures and chemicals necessary for certain tasks have been reviewed and have been promulgated by the International Commission of Microbial Safety of Foods in *Microbial Ecology of Foods,* Volume 2, published by Academic Press in 1980. Those procedures should be reviewed, and local authorities should be contacted about the permitted uses of certain chemicals because these are frequently changed as newer and more effective materials are placed on the market. Mechanical washing and rinsing, properly operated, will provide sufficient heat for a long enough time to be completely effective for disinfection. The machine must be checked at regular intervals to assure that agitation is sufficient to clean utensils of food and organic matter, and that the rinsing and drying temperatures are sufficiently high to provide disinfection. In a properly operating machine, failure to clean and/or to disinfect may result when utensils are improperly loaded or prerinsed. If soiled utensils are packed too tightly, jet sprays or blade agitated water will not reach surfaces with sufficient force to remove food or organic material. If utensils are packed so as to form air pockets and protect surfaces, neither hot water rinses nor drying heat will be effective for disinfection. These factors make it highly desirable that workers operating these machines be trained in the use of the machine and that these machines not be overloaded or improperly loaded. If the machine is in good working order and operated properly, this method is the most desirable means of cleaning utensils because it requires no intermediate handling, which may result in recontamination, and the utensils may be subjected to higher heat for a longer period of time, which helps to ensure disinfection.

The main reason for cleaning equipment and food handling areas in any food establishment, whether processing or service, is to remove food particles and soils from surfaces so that any food poisoning or spoilage microorganisms present in the environment will not have a suitable environment for growth, from which the food products can be contaminated. A second reason is to maintain a general feel and attitude of cleanliness so that higher standards of cleanliness and hygiene are encouraged among employees. Any cleaning program should be designed to fit the food materials and the processes of the establishment, because the type of detergent, pH, and temperature of the cleaners all are governed by the kind of soil to be removed from the environment. The frequency of the cleaning program depends upon the schedule of the establishment for the handling of food products. Except for the use of dishwashing machines, the cleaning processes in foodservice and retail food stores are usually manual. This is true of disinfection procedures as well as cleaning. Therefore, simple tools and a very limited variety of chemicals are usually employed.

General-purpose cleaners are usually suitable for cleaning most floors, walls, windows, cabinets, and other parts of the work areas. Such a cleaner can be found in the commercial field in containers that can be joined to a water source so that a predetermined concentration is supplied for the cleaning process. If there is any potential for back siphoning, such attachment may not be allowed locally for some water systems. Alkaline, nonionic detergents used at the recommended concentration and temperature are commonly used in dishwashing machines where washing is done at a temperature ranging from 60° C to 75° C (140–176° F), followed by a rinse with water at a temperature range of 75° C to 80° C (167–176° F), or it is done with warm water between 25° and 50° C (76–121° F) and followed by a rinse with an approved disinfectant (approved for use as a final rinse). Special cleansers such as scouring powders or grease solvents may be needed where there is a heavy residue of foods or organic residues such as may occur in sinks or some cooking utensils. Although needed in some cases, excessive abrasiveness or scoring chemicals or tools should be avoided whenever possible in order that food contact surfaces are not made rough to the point of enhancing protection of contamination. Some materials are desirable for use in manual cleaning operations because of their single-use nature. This is true of disinfection procedures as well as cleaning. Therefore simple tools and a very limited variety of chemicals are usually employed.

Cleaning processes in retail food stores are similar to food service cleaning procedures. General purpose cleaners are usually suitable for cleaning most floors, walls, windows, cabinets and other parts of the work areas. Such a cleaner can be found in the commercial field in containers that can be joined to a water source so that a predetermined concentration is supplied for the cleaning process. If there is any potential for back siphoning, such attachment may not be allowed locally for some water systems. Alkaline, nonionic detergents used at the recommended concentration and temperature are commonly used in dish washing machines where washing is done at a temperature ranging from 60° to 75° C (140–167° F), followed by a rinse with water at a temperature range of 75° to 80° C (167–176° F), or it is done with warm water between 25° and 50° C (76°–121° F) and followed by a rinse with an approved disinfectant (approved for use as a final rinse). Special cleansers such as scouring powders or grease solvents may be needed where there is a heavy residue of foods or organic residues such as may occur in sinks or some cooking utensils. Although needed in some cases, excessive abrasiveness or scoring chemicals or tools should be avoided whenever possible in order that food contact surfaces are not made rough to the point of enhancing protection of contamination. Some materials are particularly desirable for use in manual cleaning operations because of their single-use nature.

These include disposable paper towels, disposable cloths or sponges, and disposable mop heads. If disposable tools can not be used, cloths, sponges, mop heads, and nylon brushes (which dry rapidly) should be detachable so that they can be enclosed in wash bags and cleaned and heat sterilized in this manner. For all these operations it is essential that potable water be used. The use of contaminated water, particularly in the rinse phase of a cleaning operation, recontaminates the materials and/or utensils with different organisms.

The requirements for facilities and methods for washing and sanitizing dishes, utensils, glassware, flatware, and other equipment are detailed in the *Food Service Sanitation Manual,* which has been referred to frequently in this chapter.

(a) For manual washing, rinsing and sanitizing of utensils and equipment, a sink with not fewer than three compartments shall be provided and used. Sink compartments shall be large enough to permit the accommodation of the equipment and utensils, and each compartment of the sink shall be supplied with hot and cold potable running water. Fixed equipment and utensils and equipment too large to be cleaned in sink compartments shall be washed manually or cleaned through pressure spray methods.

(b) Drain boards or easily movable dish tables of adequate size shall be provided for proper handling of soiled utensils prior to washing and for cleaned utensils following sanitizing and shall be located so as not to interfere with the proper use of the dishwashing facilities.

(c) Equipment and utensils shall be preflushed or prescraped and, when necessary, presoaked to remove gross food particles and soil.

(d) Except for fixed equipment and utensils too large to be cleaned in sink compartments, manual washing, rinsing and sanitizing shall be conducted in the following sequence:

(1) Sinks shall be cleaned prior to use.

(2) Equipment and utensils shall be thoroughly washed in the first compartment with a hot detergent solution that is kept clean.

(3) Equipment and utensils shall be rinsed free of detergent and abrasives with clean water in the second compartment.

(4) Equipment and utensils shall be sanitized in the third compartment according to one of the methods included in section 5-103 (e) (1) through (4) of this ordinance.

(e) The food-contact surfaces of all equipment and utensils shall be sanitized by:

(1) Immersion for at least one-half (½) minute in clean, hot water at a temperature of at least 170° F (77° C); or

(2) Immersion for at least one minute in a clean solution containing at least 50 parts per million of available chlorine as a hypochlorite and at a temperature of at least 75° F (24° C); or

(3) Immersion for at least one minute in a clean solution containing at least 12.5 parts per million of available iodine and having a pH of not higher than 5.0 and a temperature of at least 75° F (24° C); or

(4) Immersion in a clean solution containing any other chemical sanitizing agent allowed under 21 CFR 178.1010[10] that will provide the equivalent bactericidal effect of a solution containing at least 50 parts per million of available chlorine as a hypochlorite at a temperature of at least 75° F (24° C) for one minute; or

(5) Treatment with steam free from materials or additives other than those specified in 21 CFR 173.310[10] in the case of equipment too large to sanitize by immersion, but in which steam can be confined; or

(6) Rinsing, spraying, or swabbing with a chemical sanitizing solution of at least twice the strength required for that particular sanitizing solution under section 5-103 (e)(4) of this ordinance in the case of equipment too large to sanitize by immersion.

(f) When hot water is used for sanitizing, the following facilities shall be provided and used;

(1) An integral heating device or fixture installed in, on, or under the sanitizing compartment of the sink capable of maintaining the water at a temperature of at least 170° F (77° C); and

(2) A numerically scaled indicating thermometer, accurate to $\pm 3°$ F, convenient to the sink for frequent checks of water temperature; and

(3) Dish baskets of such size and design to permit complete immersion of the tableware, kitchenware, and equipment in the hot water.

(g) When chemicals are used for sanitization, they shall not have concentrations higher than the maximum permitted under 21 CFR 178.1010[10] and a test kit or other device that accurately measures the parts per million concentration of the solution shall be provided and used.

For mechanical cleaning and sanitizing, the specified requirements include the need for proper installation and maintenance in good working order. For such operation, it is essential that proper water pressures, temperatures, and times of contact be maintained in the effective ranges. Generally speaking, the water temperatures can be higher, and the times of contact can be longer, and therefore more effective for mechanical cleaning and sanitizing because it is not necessary that the operators of the equipment be exposed to the heat of the water. Therefore, if the operator sees that the water is kept clean, the cleaning and sanitizing operation should be more efficient. The temperatures needed in different types of machines, and at different stages of the operation are detailed in Table 11.3.

Once utensils and dishes have been cleaned and disinfected or sanitized, most codes require that they be stored in such a manner that

Table 11.3.   Temperature Requirements for Dishwashing Machines

Machine Type	Temperature Requirements	
	Wash	Final Rinse
Single-tank, stationary-rack, dual-temperature	150° F (65.6° C)	180° F (82.2° C)
Single-tank, stationary-rack, single-temperature	165° F (73.9° C)	165° F (73.9° C)
Single-tank, conveyor	160° F (71.1° C)	180° F (82.2° C)
Multi-tank, conveyor*	150° F (65.6° C)	180° F (82.2° C)
Single-tank, pot, pan, and utensil washer (either stationary or moving-rack)	140° F (60° C)	180° F (82.2° C)

*Pumped rinse temperature: 160° F (71.1° C).
Data taken from *Food Service Sanitation Manual*, 1976.

they will remain sanitary. This usually means that there must be minimal handling, and that there must be clean, dry, dust-free storage facilities available. Without such storage areas, the efforts expended in cleaning and sanitizing are wasted. A partial recommendation of the manual is quoted as follows:

5-202 Storage.

(a) Cleaned and sanitized utensils and equipment shall be stored at least 6 inches above the floor in a clean, dry location in a way that protects them from contamination by splash, dust, and other means. The food-contact surfaces of fixed equipment shall also be protected from contamination. Equipment and utensils shall not be placed under exposed sewer lines or water lines except for automatic fire protection sprinkler heads that may be required by law.

(b) Utensils shall be air dried before being stored or shall be stored in a self-draining position.

The storage of single-service articles is also specified in recommendations included within the manual.

5-203 Single-service articles.

(a) Single-service articles shall be stored at least 6 inches above the floor in closed cartons or containers which protect them from contamination and shall not be placed under exposed sewer lines or water lines, except for automatic fire protection sprinkler heads that may be required by law.

(b) Single-service articles shall be handled and dispensed in a manner that prevents contamination of surfaces which may come in contact with food or with the mouth of the user.

(c) Single-service knives, forks and spoons packaged in bulk shall be inserted into holders or be wrapped by an employee who has washed his hands immediately prior to sorting or wrapping the utensils. Unless single-service knives, forks and spoons are prewrapped or prepackaged, holders shall be provided to protect these items from contamination and present the handle of the utensil to the consumer.

5-204 Prohibited storage area.

The storage of food equipment, utensils or single-service articles in toilet rooms or vestibules is prohibited.

Some ordinances specify the maximum amount of bacterial contamination that is allowable on any utensil surface. When such specifications are made, the foodservice operator will have an established criterion by which he or she can measure the sanitary operation of the business. Such specified limits also establish standards for inspectors, thus removing the necessity for judgment decisions on sanitary quality in these cases. When utensils are tested for bacterial presence and found to exceed these limits, then the establishment does not meet the sanitary code, and the point of failure in the operations schedule must be located and corrected.

It is not unusual that a health code will require a record of scheduled sanitation procedures used as well as a record of self-inspection results for specified aspects of sanitation. Such scheduled inspection procedures and records provide an excellent system for training and supervision of personnel which will result in an operation in which good sanitation practices become habit rather than extra precautions.

Steam Tables

Steam tables provide a most desirable method of maintaining food at a desired temperature for considerable periods of time following preparation. Properly functioning steam tables permit the serving of large quantities of foods at the most desirable temperature while assuring that the food will be safe from the standpoint of sanitation. To provide this benefit, it is necessary that the table operate in such a manner that all food stored in it be maintained at a temperature of not less than 140° F (60° C). Any temperature less than this will permit bacterial growth and will make a microbial incubator of the table. It is not sufficient that only the water in a steam table reach 140° F (60° C); rather, the food must be maintained at this temperature and, therefore, the water must be of a sufficiently higher temperature to assure that all parts of any food are kept at this level of heat. Regular checks on the operation of steam tables must be made to be sure of any degree of safety. The table must be of an approved design and construction, according to many ordinances, and must be kept clean at all times because of the proximity to prepared foods that will not be subjected to further cooking.

Refrigeration Equipment

Health codes require that foods subject to bacterial contamination and spoilage either be kept at a temperature of 140° F (60° C), as on a steam table, or at a temperature of 50° F (10° C) or below under refrigeration. Again, it is essential that all portions of the food reach this low temperature, and therefore, the shelf temperature of 50° F (10° C) in the refrigerator is not sufficient. Refrigerator space must be adequate for storage of foods in the establishment, without overloading or crowding the refrigerator. Food containers must not be packed tightly in a refrigerator in order that air may circulate to carry the cooling to all spaces and all foods. Without this circulation, temperature pockets will be created and some foods may stay at an elevated temperature for long enough periods to permit considerable microbial growth.

Refrigeration equipment, including walk-in cold rooms, must be constructed in such a manner that no condensation moisture, or any other liquid can drop onto any stored food. Coils located above storage spaces must be protected by a drip pan, or shield. Drains from refrigerators must be properly connected for disposal because some ordinances prohibit these from emptying onto soil or directly into waste pipes. Local regulations should be checked on this point. Floors of walk-ins should be sloped and drained, and should have a hard, smooth surface for cleaning.

It is fairly common to assume that because the purpose of refrigeration is to stop microbial growth and to prevent food spoilage, the refrigerator stays clean and does not need to be washed and disinfected. This is a false assumption and must be corrected to prevent much food contamination in any foodservice establishment. A temperature of 50° F (10° C) or below slows microbial growth and will stop much of it; however, this temperature will kill very few microorganisms and will permit the growth of many types. Many microorganisms, including many that may cause disease, can be preserved for fairly long periods under refrigeration. The presence of microbial cells in foodstuffs provides an additional amount of protection to the cell, and some very small amount of growth, even of pathogens, may occur under these conditions. Some fungi and those bacteria known as psychrophiles will grow at refrigerator temperatures, and will, over long periods of time, produce food spoilage. The growth of fungi in a refrigerator is especially troublesome due to the production of spores by the organism, which are easily scattered and will remain viable for months or even years at this temperature, contaminating foods throughout the entire time.

Some health codes specify that refrigerators shall be cleaned at least once each week, and others simply state that they shall be kept clean, sanitary, and in good repair. Cleaning of refrigeration spaces with a good detergent and with an odorless disinfectant will prevent accumulation of spilled foods and microorganisms, as well as removing accumulated odors that absorb to surfaces in such spaces. Disinfectants will require longer times for effectiveness at this lower temperature and rapid disinfection must never be assumed. Cleaning procedures should provide extra attention to cracks, crevices, seams, and shelves to remove any spilled food materials that may get into these protected places. Because of the time required for thorough, complete cleaning, it is essential that the establishment have available space for transfer of stored foods so that these may be kept under refrigeration during the cleaning process. Removal of foods to room temperature during cleaning violates the regulation requiring that such foods be kept at a temperature of 50° F (10° C) or below at all times. Removal of foods is essential for adequate cleaning, and therefore sufficient refrigeration space for transfer of foods is necessary.

Inspection of refrigerators at regular intervals, including testing of temperature, must be a regular step in sanitation procedures to avoid difficulties due to temperature variation. A refrigerator that fails to cool rapidly becomes a microbial incubator. Without specific inspection, even temporary failures or stoppages will result in considerable loss from spoilage or contamination.

PROTECTION OF FOODS

The purpose of all health codes is to assure that food and drink served to any consumer is clean, wholesome, free from spoilage, and properly prepared. To accomplish this end, foods in storage, preparation, display, holding, or serving stages must be protected from contamination and spoilage. Two types of protection have been discussed earlier in high and low temperature facilities. Specialized protective measures are required for some foods in other situations. The broad coverage of the requirements for food protection is illustrated by the following from the *Rules and Regulations for Food Service of the State of Georgia:*

290-5-14-.04 FOOD PROTECTION. Amended.

(1) All food, while being stored, prepared, displayed, served, or sold in food-service establishments, or transported between such establishments, shall be protected against contamination from dust, flies, roaches, rodents, and other vermin; unclean utensils and work surfaces; unnecessary handling; coughs and sneezes; flooding, drainage, and overhead leakage; and any other source of contamination.

(2) Conveniently located refrigeration facilities, hot food storage and display facilities, and effective insulated facilities, shall be provided as needed to assure the maintenance of all food at required temperatures during storage, preparation, display, and service. Each cold-storage facility used for the storage of perishable food in non-frozen state shall be provided with an indicating thermometer accurate to ±2° F., located in the warmest part of the facility in which food is stored, and of such type and so situated that the thermometer can be easily and readily observed for reading.

(3) Temperatures:

(a) all perishable food shall be stored at such temperatures as will protect against spoilage;

(b) all potentially hazardous food shall, except when being prepared and served, and when being displayed for service, be kept at 45° F. or below, or 140° F. or above;

(c) all potentially hazardous food, when placed on display for service, shall be kept hot or cold as required hereafter:

1. If served hot, the temperature of such food shall be kept at 140° F. or above;

2. If served cold, such food shall be:

(i) Displayed in or on a refrigerated facility which can reduce or maintain the product temperature at 45° or below; or

(ii) Prechilled to a temperature of 45° F. or below, when placed on display for service, and the food temperature shall at no time during the display period exceed 55° F.

(d) following preparation, hollandaise and other sauces may be exempt from the temperature requirements of subsection (3) (c) of 290-5-14-.04 if they are prepared from fresh ingredients and are discarded as waste within three hours after preparation. Where such sauces require eggs as an ingredient, only fresh shell eggs shall be used.

(e) frozen food shall be kept at such temperatures as to remain frozen, except when being thawed for preparation or use. Potentially hazardous frozen food shall be thawed at refrigerator temperatures of 45° F. or below; or under cool, potable running water (70° F. or below); or quick-thawed as part of the cooking process; or by any other method satisfactory to the health authority.

Other health codes may specify other measures; however, the requirement to prevent any contamination can be used by a health officer to assure the sanitary quality of any food. A few simple precautions will go far toward compliance with this regulation:

(1) Foods should not be touched by human hands, if served uncooked, or after cooking, if such contact is avoidable. If unavoidable, the workers must thoroughly wash hands prior to and frequently during the time of contact with the food.

(2) A prepared food, either in storage or ready for serving or holding, should be covered by a close-fitting clean cover that will not collect any loose dust, lint, or other extraneous matter. If the nature of the food does not permit protection of this nature, then it should be placed in an enclosed dust-free cabinet at the proper temperature.

(3) Foods in individual wrappers or containers should be dispensed directly from those wrappers or containers (e.g., milk, candies, crackers, etc.) in the presence of, or preferably by the consumer. Milk not individually packaged, must, by most codes, be dispensed from a closed, refrigerated dispenser that does not require any handling or exposure to air.

Fig. 11.1. Foodservice line with well-constructed sneeze bar to prevent customer handling of foods and to protect foods against airborne contamination.

Fig. 11.2. Improper foodservice line does not protect foods.

(4) Foods served from a buffet should be protected by maintaining foods on a steam table, or in an ice tray, at the proper temperature, and should be protected during display by a transparent shield over and in front of the food. This shield should adequately protect the food from handling by the customer and from sneezes, coughs, or other customer-originated contamination (see Figures 11.1 and 11.2).

(5) Suitable implements, such as spoons, spatulas, etc., must be provided to servers so that in no case will it be necessary for the worker to come in contact with any food during service.

(6) Multiple-service utensils for dispensing dressings, sauces, etc. should be changed and thoroughly cleaned at frequent intervals. These should never be wiped with a multiple-use cloth while filled and in use.

(7) Individual servings of food, once served, must never be served again. One possible exception is when the food is wrapped and sealed and the seal has not been broken.

(8) Any food that has contacted any surface other than the clean preparation surface, a cooking utensil or server, or serving dish should be discarded.

(9) Serving utensils should be cleaned and made sanitary between each period of use. These should never be set aside for a period of time and then reused without cleaning.

(10) Workers should be trained to handle dishes and eating utensils in a manner that avoids contact of the worker's hand with any surface that will be touched by food or the consumer's mouth. Careless handling of silverware, glasses, or cups is particularly objectionable, but is little more dangerous than contact with the surface of plates and serving dishes from the standpoint of sanitation.

STORAGE

290-5-14-.06 STORAGE. Amended.

(1) Containers of food shall be stored above the floor, on clean racks, dollies, or other clean surfaces, in such a manner as to be protected from splash and other contamination.

(2) Food not subject to further washing or cooking before serving shall be stored in such a manner as to be protected against contamination from food requiring washing or cooking.

(3) Wet storage of packaged food shall be prohibited.

The preceding storage requirements are taken from the *Rules and Regulations of the State of Georgia*. Storage of food supplies is essential in all foodservice establishments to provide an adequate volume for the demand. To have this stored supply convenient as needed, facilities must be available to maintain the foods in a clean, wholesome, sanitary condition. These facilities, therefore, should provide adequate space with the proper temperature that can be protected from dust, insects, vermin, rodents, and other contamination. The most frequent deficiency in storage facilities is too small an amount of space.

Storage rooms should provide shelves or racks above the floor (some codes require a minimum of twenty-four inches above the floor for certain foods). The shelves or racks should be washed with appropriate cleaning compounds and disinfected on frequent, regular schedules. Shelves or racks should be arranged to permit the removal and use of those supplies that have been stored for the longest period of time. Such regular rotation permits more efficient cleaning and avoids loss of supplies by reason of age.

All storage facilities should be maintained to keep supplies dry and at a suitable temperature. Storage areas must be protected from insects and rodents; however, the use of pesticides in such areas must be accomplished in a manner that will not contaminate foods. Only approved chemical treatments should be used in these facilities. Frequent stock rotation, cleaning, and disinfection will be a great help in preventing infestation. Trash and garbage should never remain in a food storage area.

The nature of a foodservice establishment results in production of much unusable waste from food, and in litter or rubbish in the form of packaging and containers. Such refuse must be placed in suitable closed containers for removal from the foodservice business. It is much preferred, and some codes require, that separate containers be used for garbage and for litter and rubbish. Clean, disinfected receptacles should be located in work areas to receive wastes, trimmings, and packaging. These receptacles should be seamless, with closely fitting lids kept closed. Plastic liners that can be discarded with the garbage and trash are most desirable, and will pay for themselves in savings realized in the time and chemicals required for cleaning. Even though such liners are used, receptacles should be washed and disinfected frequently and regularly. These containers should be removed at least once daily, as required by many ordinances; however, removal more often will avoid spoilage in the containers. Such garbage containers in food preparation areas should not be used for garbage or litter not produced in food preparation.

Extra containers must be supplied and available at locations specified by health officers; and certainly, adequate containers must be located in utensil cleanup areas. If food-waste grinders are used, they must be of a design, capacity, and installation specified by the local health authority. Such mechanical garbage disposal will greatly reduce the volume, nuisance value, and work required for removal. However, without proper equipment, installation, and operation, such a system may create other problems. Seamless containers using plastic liners will work very well from a sanitation standpoint if kept adequately clean and removed from the work areas sufficiently often. All such containers must be kept closed except for deposit of garbage.

Many ordinances require that garbage containers be cleaned and disinfected daily. Without this frequent cleaning and disinfection, garbage will sour and decay, attracting insects and rodents, even though the containers are tightly closed. These codes also require that the foodservice establishment not create a nuisance in the storage or deposit of garbage containers either outside the establishment or in a storage room. A storage room with adequate racks off the floor; with ceilings, walls, and floors of a smooth, nonabsorbent, cleanable surface, and with floor sloping to an approved drain is preferable. In such a facility, when the garbage has been removed, the containers and the room can and must be completely cleaned and disinfected. If these containers must be cleaned outside the establishment, they should be stored and cleaned on a cement slab sloped to an adequate flush type drain. Any other arrangement will create a public health hazard from the water and remaining garbage in the cleaning water which will attract insects and rodents.

Fig. 11.3. Improper solid waste disposal (*left*) and proper solid waste disposal (*right*) with packaged wastes deposited in closed receptacle.

Fig. 11.4. These illegal solid waste receptacles at foodservice establishments attract insects, rodents, dogs, cats, etc.; add to the pollution of the environment; and help spread disease.

Handling of garbage and rubbish in a foodservice establishment may be summed up by a few simple rules:

(1)   Use only good, clean, disinfected containers with tightly fitting covers.
(2)   Change containers frequently, and remove garbage from the premises frequently.
(3)   Allow no overflow of garbage or cleaning waters under any circumstances.
(4)   Leave no residues from emptying and cleaning containers.
(5)   *Keep all garbage in suitable containers and keep covers closed tightly.* See Figures 11.3 and 11.4 for examples of improper and illegal handling of solid wastes at foodservice establishments.

RODENTS AND INSECTS

The infestation of a food establishment by insects, rodents, or vermin of any kind is prohibited by public health codes. This prohibition may be stated, as in the *Georgia Rules and Regulations for Food Service:*

VERMIN CONTROL. Amended.
(1) General:
(a) effective control measures shall be utilized to prevent the presence of rodents, flies, roaches, and other vermin on the premises.
(b) the premises shall be kept in such condition as to prevent the harborage of vermin.
(2) Screening:
(a) all openings to the outer air shall effectively prevent the entrance of

insects by self-closing doors, closed windows, screening, controlled air currents, or other means;

(b) screening material shall be not less than 16-mesh to the inch or equivalent;

(c) screen doors to the outer air shall be self-closing; and screens for windows, doors, skylights, transoms, and other openings to the outer air shall be tight-fitting and free of breaks.

(3) Rodent Proofing:

(a) all openings to the outside shall be effectively protected against the entrance of rodents.

It is not possible to exclude all insects from any building to which the public has access. There are, however, effective rodent-proofing measures that can be built into these buildings. Local health and building officials should be consulted for rodent-proofing inspection. Any building used for a food industry can be maintained in such a manner that insects and rodents will not be attracted if the premises are kept clean and sanitary as required by health codes, and if food products are not exposed to these vermin.

For those insects and rodents that manage to enter, specific control and eradication measures are available. Regular, frequent inspection will detect the presence of such infestation. Detection must be followed by immediate action to eliminate the pests. Local health officials should be consulted for approval of methods to be used in pest eradication. All foods, utensils, and equipment must be protected so that they will not be chemically contaminated by pesticides used, and those methods used must not result in residual danger to food products, workers, or customers. Methods used must not leave dead pests in locations from which contamination of foods can occur. Eradication of pests by methods involving no danger to the human is possible and maintaining a sanitary establishment following such extermination is the best insurance against reinfestation.

OTHER ANIMALS

Most foodservice codes prohibit the presence of any animal other than those accompanying the blind on the premises of any food products establishment. Obviously, when animals are in the proximity of food processing, preparation, or service there is danger of contamination and such danger must be eliminated by exclusion of the animals. Some special conditions permit the exception of aquatic animals in some localities.

TOILETS AND LAVATORIES

The specifications requiring toilet and lavatory facilities may vary in different localities; however, sanitary codes are specific in the requirements that these facilities be provided. The *Food Service Sanitation Manual* recommendations are usually followed.

TOILET FACILITIES

6-401 Toilet installation.
Toilet facilities shall be installed according to law, shall be the number required by law, shall be conveniently located, and shall be accessible to employees at all times.
6-402 Toilet design.
Toilets and urinals shall be designed to be easily cleanable.
6-403 Toilet rooms.
Toilet rooms shall be completely enclosed and shall have tight-fitting, self-closing, solid doors, which shall be closed except during cleaning or maintenance, except as provided by law.
6-404 Toilet fixtures.
Toilet fixtures shall be kept clean and in good repair. A supply of toilet tissue shall be provided at each toilet at all times. Easily cleanable receptacles shall be provided for waste materials. Toilet rooms used by women shall have at least one covered waste receptacle.

LAVATORY FACILITIES

6-501 Lavatory installation.
(a) Lavatories shall be at least the number required by law, shall be installed according to law, and shall be located to permit convenient use by all employees in food preparation areas and utensil-washing areas.
(b) Lavatories shall be accessible to employees at all times.
(c) Lavatories shall also be located in or immediately adjacent to toilet rooms or vestibules. Sinks used for food preparation or for washing equipment or utensils shall not be used for handwashing.
6-502 Lavatory faucets.
Each lavatory shall be provided with hot and cold water tempered by means of a mixing valve or combination faucet. Any self-closing, slow-closing, or metering faucet shall be designed to provide a flow of water for at least 15 seconds without need to reactivate the faucet. Steam-mixing valves are prohibited.
6-503 Lavatory supplies.
A supply of hand-cleansing soap or detergent shall be available at each lavatory. A supply of sanitary towels or a hand-drying device providing heated air shall be conveniently located near each lavatory. Common towels are prohibited. If disposable towels are used, easily cleanable waste receptacles shall be conveniently located near the handwashing facilities.

6-504 Lavatory maintenance.
Lavatories, soap dispensers, hand-drying devices and all related fixtures shall be kept clean and in good repair.

The preceding requirements are necessary because of the need for adequate and sanitary disposal of human wastes, and because of the fact that the hands are probably the most common vehicle for transmission of microbial contamination to foods in preparation, and those being consumed. Maintenance of clean and attractive facilities encourages the proper use of toilet and lavatory facilities to maintain sanitary conditions within a foodservice establishment.

The maintenance of toilet facilities for the well-being and convenience of employees and customers required by these recommendations includes the obligation to keep the facilities in the proper working and sanitary condition at all times. Regular and frequent cleaning, disinfection, and deodorizing of restrooms in public places is absolutely essential for maintenance of sanitation. There can be no emergency of greater danger to a food industry than to have toilet facilities not operating effectively. Any failure to operate properly must be immediately corrected and the facility must be closed to all persons until this is accomplished. The required maintenance of restrooms is well described in the *Rules of the Texas Health Department* as included in the Appendix.

Tile or other hard, smooth surface, and nonabsorbent floors, walls, and cubicles facilitate cleaning of restrooms. The number of cleaning tools and chemical compounds available commercially for such cleaning make it extremely simple for cleaning and disinfection to be carried out with great frequency. Many of these compounds also serve as deodorizers and when used for cleaning often enough, result in an environment with a clean, sanitary odor rather than the potentially unpleasant odor of an added deodorant spray or chemical. An exhaust fan in constant operation and regular, thorough cleaning and disinfection will maintain these facilities in a pleasant, sanitary state.

Handwashing facilities, kept clean and disinfected, adjacent to toilet facilities are required by health codes, and many codes also include a posted sign stating that employees must wash hands before returning to work. For such facilities to serve the desired purpose, it is imperative that they be cleaned and disinfected with the same regularity and thoroughness used in the facility generally. Soap or detergent must be available, and it is much preferred that these be dispensed in such a manner that soil or microorganisms are not transferred from one user to the next. The actual process of hand cleaning is not always efficient, and to be truly effective must include a goodly amount of scrubbing.

Most hand soaps and detergents are not rapidly microbicidal. Instead, these cleansers serve to help loosen soil, oils, and microbial cells on the skin, so that with some scrubbing in water, these contaminants will be removed. Spreading a little soap or detergent on part of the surface and rinsing without scrubbing will not adequately clean the hands.

It is necessary that drying devices be supplied at the handwashing facility. These must be of an individual nature, either paper, cloth, or air. Common towels are prohibited. If towels of any nature are used, the supply must be adequate for the needs of the patrons and employees of the establishment, and if disposable towels are used, receptacles are necessary for discarding after usage.

Many health codes require that new foodservice establishments install handwashing facilities of equivalent adequacy to those just described at convenient places in food preparation areas. Such lavatories, exclusively for handwashing, are probably as valuable as those in toilet facilities, from the standpoint of good sanitation. As workers handle foods and utensils in food preparation areas, hands become soiled. The worker may unconsciously run hands across the face, hair, or clothing, or when hands are soiled and sticky from foods, may pick up the nearest handling cloth to wipe the hands. When such events occur, the hands are contaminated by microorganisms on the skin, body, clothing, hair, or in the respiratory system. The longer the hands remain unwashed, the more microorganisms accumulate and grow on the skin and the more they may be transferred to foods.

Or, the microorganisms may come from the food being handled. For example, a worker may remove poultry from the refrigerator, including a chicken carcass contaminated with *Salmonella*. The worker prepares all the chickens, spreading the *Salmonella* cells to all, and places them in the oven to cook. The contamination on the chicken meat is killed by proper cooking. However, if the worker now turns to preparing already cooked foods (breads, pastries, salads, etc.) the *Salmonella* cells remaining on the hands will be transferred to these foods and then to the customer.

If handwashing facilities are convenient, adjacent to the preparation area, and if workers are trained to wash hands thoroughly between preparatory steps, these cases of bacterial contamination transferred from the worker's body, or from one food to another, will be stopped before they occur. Frequent handwashing by food workers keeps the hands of the workers from accumulating contaminating microorganisms from articles handled, reduces the multiplication of organisms on the skin, and keeps the personnel more conscious of good personal hygiene in the food preparation areas. Some codes, in effect, require a number and location of lavatories that will not interfere in any way with their use by workers. This means essentially that there must be sufficient numbers of lavatories in locations that will encourage good personal hygiene for the protection of the public health.

CLEANING OF FOOD ESTABLISHMENTS

In addition to specifications for the cleaning of utensils, equipment, and dishes, there are many codes that also govern cleaning and maintenance of all food industry premises. These may vary from a general requirement that all parts of the establishment be kept neat, clean, and free of clutter, to specific methods of cleaning, including prohibition of certain methods or cleaning agents. In practically all codes, it is a requirement that dustless methods of cleaning floors, walls, and cabinet surfaces be used. This means that vacuum sweeping or wet cleaning must be used on these areas, and even these should be done only when precautions have been taken to prevent exposure of foods, utensils, and equipment. Cleaning of floors should be done at least once daily, or at more frequent intervals if necessary to maintain sanitation.

For keeping the sanitary standards of a food industry, it is required that adequate facilities such as sinks, hose connections, brooms, brushes, mops, detergents, trash containers, and any needed cleaning equipment be supplied for use by well-trained and supervised personnel. In no case should the operator of the establishment permit neglect of routine housekeeping cleaning operations. Other sanitary precautions will be negated by such neglect. Mere removal of obvious trash, garbage, and litter is not sufficient. The entire premises must be cleaned and made sanitary by regular, adequate, and thorough cleaning to make sanitation procedures in other parts of the operation meaningful.

EMPLOYEE DRESS

It is not generally required that employees of foodservice establishments wear special clothing other than normal attire in carrying out their duties, but most codes do state that the clothing and person of workers be neat and as clean as practicable in consideration of their specific duties. This means, at a minimum, that workers appear for work clean and in fresh, neat clothing. Generally this clothing is required to be washable; a requirement intended to encourage keeping the clothing clean. The *Food Service Sanitation Manual* recommendation states "(a) The outer clothing of all employees shall be clean. (b) Employees shall use effective hair restraints to prevent the contamination of food or

Fig. 11.5. Spacious, uncluttered dressing- and washrooms permit employees of food-service establishments to maintain personal cleanliness and will encourage efforts toward sanitation in the entire operation. Dressing- and washrooms should offer closed lockers, showers, toilet facilities, handwashing facilities with adequate soap, drying facilities, and trash depositories.

food-contact surfaces." The use of such restraints will generally serve a two-fold purpose. First, loose hairs are prevented from falling onto foods; and second, personnel are in some measure hindered from passing the hands through or over the hair while at work. It is a simple matter for a microbiologist to demonstrate to anyone the number of microorganisms that are likely to be found on any single hair, regardless of how recently it has been shampooed. These microorganisms will dangerously contaminate foods directly, and are easily spread from hair to hands to foods. Foodservice managers should see that this requirement is adequately enforced, and that the hair restraint is truly effective in preventing these contamination pathways from becoming prominent.

When possible, it is desirable that special clothing be worn at work in foodservice, rather than the usual street clothing. When this is done, dressing room facilities must be provided. Such dressing room facilities are required to be separate from toilet facilities and to be separate from all food storage, preparation, and service areas. Dressing rooms must be of adequate size for the number of persons using them, and must provide facilities for storage of both street and work clothing in separate locations (see Figure 11.5). These spaces must be kept as clean and sanitary as any other portion of the foodservice establishment.

EMPLOYEE CONDUCT

In addition to their dress, employees also affect the sanitary state of the foodservice establishment in other ways. The *Rules and Regulations for Food Service of the State of Georgia* includes the following section dealing with personnel.

290-5-14-.10 PERSONNEL. Amended.

(1) Health and Disease Control:

(a) disease control

1. No person while affected with a disease in a communicable form, or while a carrier of such disease, or while afflicted with boils, infected wounds, or an acute infection, shall work in a food-service establishment in an area and capacity in which there is a likelihood of transmission of disease to patrons or to fellow employees, either through direct contact or through the contamination of food or food-contact surfaces with pathogenic organisms. No person shall be employed in such an area and capacity in a food-service establishment.

(b) reporting

1. The manager or person in charge of the establishment shall notify the health authority when any employee of a food-service establishment is known or suspected of having a disease in a communicable form.

(c) cleanliness

1. Employees shall maintain a high degree of personal cleanliness and shall conform to good hygienic practices during all working periods.

2. Handwashing

(i) All employees shall thoroughly wash their hands and arms with soap and warm water before starting work, and shall wash hands during work hours as often as may be required to remove soil and contamination, as well as after visiting the toilet room.

(ii) Employees shall keep their fingernails clean and neatly trimmed.

(d) clothing

1. The outer garments of all persons, including dishwashers, engaged in handling food or food-contact surfaces shall be reasonably clean.

2. Hair nets, headbands, caps, or other effective hair restraints shall be used by employees engaged in the preparation and service of food.

(e) tobacco

1. Employees shall not use tobacco in any form while engaged in food preparation or service, or while in equipment and utensil washing or food-preparation areas.

Some codes also prohibit the use of any part of a food products establishment for living or sleeping by personnel. Generally, those acts by a person that can contaminate foods, utensils, equipment, or the establishment itself with filth from the worker's body or from things used personally by the workers are prohibited. If each worker is trained to those personal behavior traits that will result in the service of foods attractive to that worker and to his or her co-workers, these acts are acceptable from the standpoint of sanitary practices.

A most difficult area to control is the presence, employment, or continued employment of persons with infectious disease. Some infectious diseases, including acute respiratory infections and skin infections where face or hands are affected, are easily recognized by both the infected person and his or her untrained supervisor. These persons are easily prevented from spread of pathogenic organisms by change of duties or requiring them to remain off duty until the infection is cleared by treatment. Many contagious, infectious diseases, however, are not readily recognized and may be spread before they are detected. Codes make it unlawful for persons with such diseases to work in foodservice establishments even though the disease is of this type and both the afflicted person and the employer are unaware of the condition. Preventing a situation of this type necessitates frequent health care and examination of persons employed in foodservice establishments.

The requirement of health examinations at intervals of six months or one year will possibly detect only low grade, chronic infections and will do little to provide protection against the sporadic, short-lived, acute infections to which most people are susceptible, and which in many cases are the most readily transmissible. For this reason it is well for supervisors and employers to encourage all employees in foodservice industries to report possible infections, and to encourage employees to seek early medical care when infection is suspected. In addition to these precautions, the training of personnel in good personal hygiene will help to prevent spread of these difficult-to-detect diseases.

The requirement for health examination is frequently standard in health codes for food industry workers. This requirement typically includes an examination prior to beginning work and at periodic intervals thereafter. Such examinations may include tests that will detect chronic lung diseases such as tuberculosis and other chronic respiratory diseases, skin infections, and if adequate histories are given by the worker, other types of infectious disease. Blood tests are usually included that may detect syphilis infections as well as some other diseases. In recent times, the increasing incidence of AIDS has alarmed public health officials, and has created a demand that health cards and examinations be required. The requirements for such examinations may be justified in that they serve to reduce the spead of some infections. However, to realistically increase protection against the spread of pathogenic organisms of all kinds, the cooperation of food industry workers, their supervisors, and their employers is required. Most such health certificate examinations will not detect carrier states, nor will the periodicity of the examinations permit detection of many illnesses that have short incubation periods, and do not continue for long periods of time. Any suspicion of infectious diseases must be promptly followed through by thorough medical tests. Any reported illness of customers after patronage in a foodservice establishment should be reported to public health officials and should be thoroughly and completely investigated. Only by such precautions as these can infectious disease spread be adequately controlled in food industries.

SPECIAL REQUIREMENTS

Inspections

Most public health laws for control of food industries include specification for periodic inspection of the entire operation by public health officers. This specification includes the authority to enter and observe the operations in any room or building used in a foodservice business, at any time, without prior notification. The purpose of this is not to intimidate or trap operators of foodservice establishments; it is rather to make it possible to observe the actual, regular operation of all phases of the establishment's activities. Some ordinances require regular self-inspection by the food establishment operator, and require that a record of all such inspections be kept. Whether or not this self-inspection is required, it is good business practice for the operator to seek a copy of an official inspection report blank, and to periodically check the operation of the business.

Official or self-inspections of these businesses are a matter of protection for the business as well as for the customers who are served. Failure for any period of time in the sanitation procedures of an enterprise dealing in foodservice is likely to result in disease in the customer and ultimately in various financial and legal difficulties for the operator of the business. Operators of food establishments protect themselves and their businesses by carrying out rigid self-inspections and by complete cooperation with public health inspectors and officials.

Poisons and Toxic Materials

It is unlawful to keep poisons or toxic chemicals on the premises of a foodservice establishment, other than those required to maintain sanitation. When sanitation requires the use of such poisons, the containers must be clearly labeled; and these chemicals must be kept in an area away from foods or food storage. Extreme caution in use and completely separate storage for such materials is simply good safety practice.

Precautions to be Taken with Certain Foods

Some foods, by their nature, require special precautions in production, handling, shipping, and storing. Foods of this type include eggs, meats, poultry, seafoods, milk, and milk products. The operator of a foodservice business will avoid trouble with these foods by personally

(a) reading and understanding local codes regulating such foods, (b) purchasing only from suppliers specifically approved by local health officials, and (c) refusing delivery of any supplies that appear to be of questionable condition in any way. Local health officials can be of greatest help to the operator in those situations involving these special foods and in giving information and advice. Time spent in checking on regulations, approved sources, and quality of foods supplied is the best safeguard for assuring good sanitary quality in these foods.

SPECIAL FOODSERVICE OPERATIONS

Mobile Foodservice

Some localities permit the operation of mobile foodservice units or pushcarts. Where this is permitted, it is essential that the regulatory authorities promulgate special regulations for the control of sanitary conditions of foodservice. The *Food Service Sanitation Manual* recommends:

8-101 General.

Mobile food units or pushcarts shall comply with the requirements of this chapter, except as otherwise provided in this paragraph and in section 8-102 of this ordinance. The regulatory authority may impose additional requirements to protect against health hazards related to the conduct of the food service establishment as a mobile operation, may prohibit the sale of some or all potentially hazardous food, and when no health hazard will result, may waive or modify requirements of this chapter relating to physical facilities, except those requirements of sections . . . of this ordinance.

8-102 Restricted Operation.

Mobile food units or pushcarts serving only food prepared, packaged in individual servings, transported and stored under conditions meeting the requirements of this ordinance, or beverages that are not potentially hazardous and are dispensed from covered urns or other protected equipment, need not comply with the requirements of this ordinance pertaining to the necessity of water and sewage systems nor to those requirements pertaining to the cleaning and sanitization of equipment and utensils if the required equipment for cleaning and sanitization exists at the commissary. However, frankfurters may be prepared and served from these units or pushcarts.

8-103 Single-service articles.

Mobile food units or pushcarts shall provide only single-service articles for use by the consumer.

8-104 Water system.

A mobile food unit requiring a water system shall have a potable water system under pressure. The system shall be of sufficient capacity to furnish enough hot and cold water for food preparation, utensil cleaning and sanitizing, and handwashing, in accordance with the requirements of this ordinance. The water inlet shall be located so that it will not be contaminated by waste discharge, road dust, oil, or grease, and it shall be kept capped unless being filled. The water inlet shall be provided with a transition connection of a size or type that will prevent its use for any other service. All water distribution pipes or tubing shall be constructed and installed in accordance with the requirements of this ordinance.

8-105 Waste Retention.

If liquid waste results from operation of a mobile food unit, the waste shall be stored in a permanently installed retention tank that is of at least 15 percent larger capacity than the water supply tank. Liquid waste shall not be discharged from the retention tank when the mobile food unit is in motion. All connections on the vehicle for servicing mobile food unit waste disposal facilities shall be of a different size or type than those used for supplying potable water to the mobile food unit. The waste connection shall be located lower than the water inlet connection to preclude contamination of the potable water system.

Foods served from mobile units or pushcarts are subject to the same kinds and degrees of contamination that threaten other foodservice operations. For this reason, then, it is essential that the consumers be offered the same kinds of protection as in other foodservice operations. Some kinds of pushcart operations cannot have all the safeguards provided by fixed foodservice operations, and these must be strictly limited to sale or service of the kinds of foods that are authorized in order that the consumers be protected. When these mobile units operate from commissaries, then it is necessary that additional regulations be in force. For that reason, the *Food Service Sanitation Manual* recommends the following for the commissary:

8-201 Base of Operations.

(a) Mobile food units or pushcarts shall operate from a commissary or other fixed food service establishment and shall report at least daily to such location for all supplies and for all cleaning and servicing operations.

(b) The commissary or other fixed food service establishment used as a base of operation for mobile food units or pushcarts shall be constructed and operated in compliance with the requirements of this ordinance.

The temporary foodservice establishment may offer additional avenues for contamination of foods, or for transmission of potential infectious organisms to the consumers. For this reason, there must be additional and somewhat specific regulations for these temporary operations. The reasoning behind this can be recognized in the general requirements quoted here for temporary foodservice:

9-101 General.
A temporary food service establishment shall comply with the requirements of this ordinance, except as otherwise provided in this chapter. The regulatory authority may impose additional requirements to protect against health hazards related to the conduct of the temporary food service establishment, may prohibit the sale of some or all potentially hazardous foods, and when no health hazard will result, may waive or modify requirements of this ordinance.

In most cases the dispensing of potentially hazardous foods from temporary establishments is rather severely limited, and the recommendations state that most temporary operations should be limited to dispensing only prepackaged and preprepared foods.

BIBLIOGRAPHY

CITY OF DALLAS. 1977. Ordinance #15578 amending chapter 17, Food establishments and drugs of the *Dallas City Code* as amended. Dallas, TX.

CITY OF HOUSTON. 1978. *The Food Ordinance.* An ordinance amending Article II of chapter 19 of the *Code of ordinances of the City of Houston regulating food establishments.* Houston, TX.

INTERNATIONAL COMMITTEE ON THE MICROBIOLOGICAL SAFETY OF FOODS. 1980. Cleaning, Disinfection, and Hygiene. In *Microbial ecology of foods.* vol. 2. New York: Academic Press.

MINOR, J. J. 1983. *Sanitation, safety, and environmental standards,* vol. 2. Westport, CT: AVI Publishing Co.

STATE OF GEORGIA. 1980. *Rules and regulations for food service* (chapter 290-5-14). Atlanta, GA.

TEXAS DEPARTMENT OF HEALTH. 1978. *Rules on food service sanitation.* Austin, TX.

U.S. DEPARTMENT OF HEALTH, EDUCATION AND WELFARE. *Food service sanitation manual.* DHEW, Public Health Service, Food and Drug Administration. Publication no. (FDA) 78-2081.

12

Sanitation in Retail Food Stores and Transportation

RETAIL FOOD STORE SANITATION

In the past there have been many major differences in the requirements and practices of sanitary procedures in foodservice establishments and in retail food stores. More recently, however, there have been so many changes in the operation of businesses involving foods that some of the differences have been gradually eliminated or at least shaded in degree. One example is that many retail food stores now operate delicatessen departments that operate essentially as take-out foodservice operations. Such an operation is found in many large supermarkets that have, as a service to their customers, a salad bar where fresh salads are sold by the pound, after they are selected and mixed by the customer.

Even before such operations were so common, the National Association of Retail Grocers had developed and prepared, in conjunction with the U.S. Department of Agriculture, a program for sanitation in retail food stores that is an excellent guide for the sanitary operation of a business of this kind. This program enables a food store to meet the local requirements for sale of wholesome, sanitary foods, and assures the customers of the supermarket that adheres to the program that the products they purchase have been carefully and properly handled, and are reaching the consumer in the best possible condition. To accomplish this goal, the purposes of a sanitation program in a supermarket are to present an attractive image to the potential customer by cleanliness, sanitation, and good housekeeping. Such an image would eliminate disordered shelves and display cases, unattractive and odorous meat counters, dirty checkout stations, and untidy employees. Presentation of this image helps to protect the public health, and saves the business and the consumer money in that foods do not spoil and do not have to be thrown out.

The materials and the practices for carrying out sanitation in any supermarket do not differ from those in a foodservice establishment. If a chemical is a good disinfectant in a foodservice operation, it is a good disinfectant in a retail food market operation. The application and the schedules of the sanitation program may vary somewhat in the retail supermarket because of the greater variety of products that must be available to the customer and the manner in which these products must be displayed and made available to the customer.

Of tremendous concern to the retail supermarket manager is the condition and appearance of the meat market. Because the operation can appear to be messy and unclean, extra effort must be made to ensure that the operation is exceptionally clean, and that temperatures are maintained exactly as required. Unless this is done, meats will be contaminated, and if not spoiled in the store, will spoil soon after removal to the customer's home. Even if spoilage does not occur, odors will develop within the store, and within a very short time the odor will so permeate the environment that it will be most difficult to remove. The National Association of Retail Grocers of the United States (NARGUS) recommends in its sanitation program that very specific cleaning and sanitizing programs be established and carried out in the meat department of supermarkets. These programs include the inspection of delivery vehicles, storage facilities, and preparation areas. If meats are to be pre-cut and packaged, then this must be done on a schedule so that the meats do not sit in the store for long periods of time after packaging. If this occurs, the meats will turn dark where exposed to the light, and will not appear attractive to the customer when seen in the display case. Some retail grocers who are not entirely conscientious may repackage such meats, opening the package and turning the meat over so that a different side is exposed to light, and the darkened side is hidden. When you purchase a package of meat like this, you have just learned much about your retail grocer. If this practice is common and occurs with great frequency in a particular store, it is sufficient reason for the customer to change markets, and the store will rapidly feel the loss of business.

Many retail supermarkets now operate bakery departments as a part of the total presentation of foods to customers. Where this is done, there is a specific need for special sanitary practices just as there is in the foodservice operation where bakery products are produced. The National Association of Retail Grocers specifies the construction and quality of facilities for bakery operation, as well as certain routines and practices that must be followed for a good sanitary program. The bakery program must include procedures to be followed for receiving goods in excellent condition; storage of supplies and ingredients; preparation and mixing of bakery products and fillings; baking; storage of cooked bakery goods, both those that are potentially hazardous, and those that are not; display of baked goods; and packaging of baked goods. Each of these steps is described and specified in the recommendations of the NARGUS committee.

The bakery department of a retail food operation may be one of the most critical areas for sanitation because of the needs for specialized food storage facilities. The NARGUS committee states that the first-in–first-out system of product storage and rotation must be used for all products. This system will reduce many of the attractants for insect and rodent infestation in this area, and if proper cleaning is done on the schedules suggested, a sanitary program should be maintained in these facilities. Other storage requirements do not vary greatly from those of other foodservice establishments.

Like the meat market, the dairy department is a critical area for the maintenance of cleanliness and sanitation. Because the food source is animal, and there is a likelihood of it containing pathogenic microorganisms when it comes into the store, the food is considered to be a hazardous food, and sanitation is absolutely essential. Because of the nature of the foods, and the manner of packaging these foods, cleanliness is absolutely essential to avoid the appearance of dirtiness, and to avoid unpleasant odors in the entire facility. The nature of the foods requires that even the display areas be refrigerated, and the committee suggests that the dairy display sections be cleaned and sanitized weekly, except for the milk areas, which are to be cleaned and sanitized daily.

It is important that the retailer inspect and closely monitor the delivery of dairy products to the store. If the foods are transported in dirty trucks that are improperly refrigerated, the retailer is going to have many and frequent complaints from the customers purchasing the products. Any damage done to the products before or during delivery cannot be corrected by any sanitary precautions or programs in the retail store. Damaged cartons or packages should be refused, and within the store, any such packages should be removed and discarded in the proper way. Temperature control within the safe ranges must be maintained in the receiving, storage, and display areas for dairy foods. In storage or refrigerated spaces assigned to dairy foods, no other materials or foods should be stored.

A department that has both the potential for bringing in pests and microbial contaminants, and for creating solid waste problems adding to the housekeeping difficulty of the area, is the produce department.

Ample space and workbench areas that are cleaned and sanitized daily must be provided for handling the produce coming in, and for checking, trimming, and re-displaying the produce that was previously delivered. It is essential that all produce being displayed be checked daily in order to trim and remove any decaying or damaged product, and to keep the produce attractive to the customers. Ample container space for waste must be provided because the daily trimmings, and those that are done at delivery will amount to large amounts of solid waste. Those items needing washing must be washed in sinks or on adequately drained workbenches that are kept clean and sanitized at frequent intervals. Because the produce has come in from the producing farms by way of the wholesaler, much of it will carry soil, insects, and potentially spoiling or disease-causing microorganisms. When these contact surfaces in the retail store, contamination occurs, and must be controlled if foods are not to be spoiled beyond use.

Another critical department in the retail food store is the frozen food department, which is present in essentially all such operations, because a major frozen food item is frozen dairy products. It is essential that the display areas be adequate so that foods are not displayed above approved fill lines, and products are never allowed to even slightly thaw. It is recommended that frozen food display cases be cleaned and sanitized at least quarterly. It is not uncommon, however, to observe cases that contain large cakes of ice from partially thawed food containers that seem to remain for many months at a time. It is not safe to purchase food items from such frozen food display cases. As in the case of other items being transported and delivered to the store, it is necessary that the retailer inspect and regularly observe the cleanliness and the temperature records of trucks that deliver these products. Products should be moved directly from the truck into a freezer box for either storage or display, immediately upon delivery. Adequate space should be provided so that the first-in-first-out rotation for display and storage can be practiced.

Cans and paper cartons holding staple groceries make it easier to manage these products for display and storage than to manage other items like meat, dairy products, and produce. Even in this department, though, it is important to practice first-in-first-out display and storage. It is necessary that potentially hazardous materials, for example, insecticides and toxic chemicals, be displayed and stored in separate areas away from foods that may be contaminated. Dented cans, and torn cartons and packages must be regularly and rapidly removed from display shelves and storage areas. These present a danger of spreading contamination to other products, and under no circumstances should they ever be sold to customers.

Good general housekeeping must be practiced in the grocery department, including cleaning and sanitizing display shelves at regular intervals; sweeping and mopping floors, walls, and ceilings on a regular basis; and cleaning and sanitizing storage racks and skids at regular intervals, although these intervals may be somewhat longer than those for general cleaning. Shelves and displayed products should be dusted and straightened when restocking in order to keep the displays orderly and attractive to the customers.

The prepared food, delicatessen, or restaurant departments will be regulated by essentially the same requirements as any other foodservice establishment. In this area, it is essential that the correct temperatures (either below 40° F or above 140° F (4°–60° C) be maintained. Deli cases should have permanent thermometers installed, and these should be visually checked at least every morning and every evening. It is preferable if the cases have installed permanent recording thermometers so that in case of failure, the exact time and degree of temperature change can be detected. Because the foods in the deli will not be further cooked, it is absolutely necessary that the workers handle the foods properly, and that hands should not touch the foods when serving them. All cases, display areas, counters, containers, trays, pans, and other utensils must be cleaned and sanitized after each use, or each contact with foods. Refrigerated display cases should be cleaned and sanitized at weekly intervals, and hot food display cases should be cleaned and sanitized daily. Floors should be cleaned and sanitized daily, and other surfaces, including walls and ceilings, at less frequent intervals.

A most important area from the standpoints of cleanliness, sanitation, odors, and neatness in the retail food operation is the seafood department. Some of the special requirements for seafood handling, storage, display, and sale will be discussed in more detail in the next chapter. Suffice it to say at this time that consistency of carrying out any sanitary procedures in other departments should be even more strongly emphasized and used in this department. Obviously, temperature controls are most critical, and it is absolutely essential that the paths of handling cooked and raw seafoods never cross within the establishment.

Within the retail food supermarket, management of pest control does not differ greatly from that in any other food establishment. It is still necessary that trash and solid wastes be removed regularly, and that receptacles be kept clean, available, and sanitized. It is still necessary that the use of insecticides or pesticides be instituted under very strict safety standards, and at regularly scheduled intervals. It is necessary, as in the case of foodservice establishments, that the use of pesticides is not allowed to contaminate any food product in any location, and that residues remaining following pesticide use not be in locations from which foods may be contaminated. The regulation that does not permit any animal except seeing eye dogs on the premises of a retail food market is again necessary.

In the *Retail Food Store Sanitation. A Total Store Concept* publication of NARGUS, checklists are provided for the use of the manager of retail food market operations that should be most helpful in establishing and maintaining an adequate sanitation program for that business.

SANITATION IN FOOD TRANSPORTATION

Food that is produced and processed under satisfactory environmental conditions must still survive transportation to the retail food establishment in satisfactory, and hopefully sanitary form. The use of clean and satisfactory vehicles for transport has been a difficult problem historically for all concerned with the problem. One of the difficulties has been in the assignment of responsibility for the condition of the transport vehicle between the shipper, the rail or truck line, and the previous user of the vehicle. When no responsibility is assigned or accepted, food products may be shipped in vehicles totally unsuited by construction or cleaning for the purpose, and the foods have been contaminated, infested with pests, or otherwise made unfit for human consumption. The Food, Drug, and Cosmetic Act actually makes all parties responsible. This fact was emphasized in 1974 when one of the United States' largest railroads, and one of its largest food manufacturing concerns were charged in an action by the FDA with unlawful contamination and interstate shipment of contaminated foods. The food processor claimed that the railroad had furnished an unsuitable car from the standpoint of cleanliness, and that the processor had loaded the car with food of a satisfactory sanitary condition. In this case the processor was not found to be at fault, but the carrier was responsible for the contamination present in the car when it was leased to the processor. Obviously, a carrier, railroad, or trucking company cannot be held responsible for food products contaminated by the processor or producer before shipment.

In 1974, the FDA organized a Railcar Sanitation Action Committee with members from food industries, railroads, and regulatory agencies to find effective methods for preventing the use of unsatisfactory and unsanitary shipping cars. One of this committee's best recommendations was that rail cars used for shipping food products be used only for that purpose and no others. The committee established lists of products that could be shipped in these cars, and any items not included in the list were not allowed. In 1976, a recommendation of the committee allowed the FDA to expand its control of railcars and trucks used to ship food, and instituted a system of records, now computerized, that can be used to trace the history of any shipping vehicle so that materials carried and shippers can be identified.

When a vehicle is leased for shipment of food products, or when an industry-owned vehicle is to be used for shipping, it is essential that the vehicle be left clean and in good condition when it is unloaded. This can be assured by inspection at three times: at arrival, during unloading, and after unloading. Such inspection must include the requirement that any damaged or spoiled products be removed, discarded properly, and the potential for contamination from these products be eliminated. Often evidence of insect or pest contamination will not be apparent during the process of unloading, but becomes so when the vehicle is empty. When such evidence is discovered, it is a simple matter still to prevent the contaminated or damaged goods from entering the facilities of the receiver. In all cases, however, it is essential that all traces and residues of foods be removed from the vehicle, leaving it completely clean, and when applicable, sanitary for the next shipper to use.

Most transport vehicles, either rail or truck, were not designed for the sanitary shipment of food articles. One of the tasks of committees dealing with regulation of transportation of food supplies has been to establish certain requirements and criteria for such vehicles so that they can be made sanitary for the shipment of foods. If these vehicles are then used only for the transport of foods, it becomes much easier to manage the consistent sanitation of materials shipped. The openings of such shipping cars should be designed so that they can be sealed to the entrance of insects and dust. Crevices, double floors, and walls where insects can hide and breed are undesirable. Walls and ceilings should be free of dust catchers and/or loose or flaked paint. Cleaning of shipping cars includes removal of trash and solid wastes, including containers as well as food residues, and proper disposal of these materials. Air vents, filters, and so on should be properly cleaned and if desirable, sanitized. Fumigation should be carried out only after removal of rubbish, and insecticides should be applied to seams, cracks, and crevices. Once insecticide has been applied, it is then well to line the entire wall with cardboard to form a physical barrier between the food products and the insecticide residue. Damaged cartons or containers of foods should never be loaded into rail or truck carriers. Loading should be done in a manner that will prevent shifting of cargoes, and avoid damage to the foods. A rail car or a truck that is to carry frozen or refrigerated foods should never be loaded until it has been precooled to the desired temperature.

BIBLIOGRAPHY

THE NATIONAL ASSOCIATION OF RETAIL GROCERS. 1978. *Retail food store sanitation. A total store concept.* Developed by the Joint Committee of the United States Department of Agriculture and the National Association of Retail Grocers. Washington, DC.

TROLLER, J. A.   1983.   *Sanitation in food processing.* New York: Academic Press.

13

Seafood Sanitation

Because of their natural habitat, and because of the fact that seafoods are so much better when fresh, there are certain very specialized requirements for seafood sanitation. In the latter years of the nineteenth century and early years of the 1900s, there were several outbreaks of enteric infection and typhoid fever that were clearly associated with consumption of raw or improperly cooked seafoods. In one such outbreak, at least 120 cases of typhoid were traced to infected fishermen who harvested oysters in New York. Such outbreaks spawned concern in the U.S. Public Health Service about the safety of seafoods. They called a conference of health officials and officials of the seafood industry in 1925 that resulted in the establishment of the National Shellfish Sanitation Program, which published its first *Manual of Operations* that same year. This program has continued and been expanded since that time, and its efficient operation has influenced all segments of seafood sanitation.

In the United States, we have been most concerned with the sanitary condition of shellfish, because these were the seafoods most often consumed raw. In other parts of the world, this was not always the case, and the sanitary condition of other seafoods were equally important. Proper cooking of seafoods inactivates or kills any potentially pathogenic microorganisms that may be present; and most seafoods at harvest will be contaminated with some microorganisms of this type. Common inhabitants of seawater and in some localities freshwater are members of the bacterial genus *Vibrio,* many of which are consistently pathogenic, and some of which are opportunistically pathogenic, dependent upon the dose ingested. In the Orient, where the practice of consuming raw fish preparations is common, infections caused by *V. parahaemolyticus* are relatively common. In the United States, for many years we assumed that our seafoods were not contaminated by this organism, because we did not see the infection. We were incorrect. We now know that this species as well as other *Vibrio* sp. inhabit our coastal waters as natural inhabitants, and unless proper precautions are taken, that is, proper cooking and sanitation, they can cause infection in humans.

Shellfish transmission of infectious disease is not only a concern in the United States; it is a concern worldwide. Because of this, the World Health Organization, from Geneva, in 1978, published a laboratory guide to shellfish hygiene. This guide describes the likelihood of shellfish contamination by pathogenic organisms; the objectives of surveillance of the shellfish environment, and the harvested shellfish for sanitary conditions; and the methods for laboratory examinations and standard establishment to control this important food industry. The Food and Drug Administration, in its *FDA Consumer,* has published frequent articles dealing with the conditions that permit the transmission of disease by contaminated shellfish. In this process, the public and industry are alerted to possibilities of disease outbreak, and what foods and practices must be surveyed to control such outbreaks.

An excellent general discussion of the nature and needs of sanitation in regard to fish, including crustacea and shellfish, is found in Volume 2 of *Microbial Ecology of Foods,* published by Academic Press in 1980. In that work, definitions are given for the different kinds of animals that we class as seafoods. These include fish, meaning finfish; crustacea, which includes lobster, crab, shrimp, prawns, and other organisms that have chitinous shells; and finally, shellfish or mollusks, which includes clams, scallops, oysters, mussels, and organisms that are generally sessile, and that feed primarily by filtering the water in which they are grown. It is this filtering feeding system that creates some of the problems associated with the consumption of raw shellfish worldwide. Depending upon the species, the water temperature, and other factors, certain shellfish may filter as much as 200 to 300 gallons of water per day. This amount of water may well contain billions of microbial cells, and if these are concentrated within the body of the mollusk, there may be many infectious doses of microorganisms in the single animal. Many of the shellfish live in shallow, brackish water, and are particularly at home in saltwater that is diluted by some incoming freshwater. This then provides the environment for the presence of large microbial populations in water brought in by freshwater streams, or stirred by tidal movement of water in the shallow estuaries, as well as dissolved chemicals that may be carried in the streams, or stirred up from the sediments in the bays.

The diseases that may be caused by consumption of raw or improperly cooked shellfish are of two types: infectious diseases or biointoxications. The infectious diseases may be caused by viruses, bacteria, or parasitic animal pathogens. One disease that has been recognized for a longer period of time, resulting from consumption of raw shellfish, is infectious hepatitis or Hepatitis A. Others very commonly transmitted by raw shellfish are typhoid, salmonellosis, polio, rotavirus, *Vibrio parahaemolyticus* infection, other *Vibrio* infections, *Clostridium botulinum* intoxication, and the paralytic seafood poisoning caused by the Red Tide organisms. Fresh, raw seafoods are not infrequently incriminated in the transmission of these diseases, and when seafoods are shipped any distance from the harvest area, even under refrigeration, the shelf life of the food is dramatically shortened because of the multiplication of bacteria in the seafood. Such shipment and storage for availability to the general consuming public constitutes one of the major problems of the seafood industry today. Of course, viruses are not replicated in the shipped or stored seafoods, but the filter-feeding nature of the mollusk makes it possible that many virus particles have been concentrated in the animal before harvest, and shipment and storage under refrigeration, even in the frozen state, will not inactivate many of the viruses present.

The *FDA Consumer* has reported that there are a number of types of viruses that may contaminate shellfish if they are exposed to sewage effluents. These include the polioviruses, Coxsackieviruses, echoviruses, enteroviruses, Hepatitis A, reoviruses, rotaviruses, parvoviruses, adenoviruses, and papovaviruses. Among the bacteria that are frequently isolated from freshly harvested shellfish are *Vibrio* sp., *Escherichia* sp., *Shigella* sp., *Salmonella* sp., *Pseudomonas* sp., *Aeromonas, Staphylococcus, Proteus* sp., *Yersinia* sp., *Campylobacter* sp., and others. Attempts have been made, and standards are set for harvest areas, based on the presence/or numbers of certain of the fecal indicator organisms for water. In these practices, the numbers of indicator organisms in the water, as well as the numbers of these organisms isolated from the meat of the shellfish have been used to establish the suitability of the shellfish for human consumption. For many of the more common bacterial and viral diseases, these bases for shellfish harvest approval have worked fairly well. In the case, however, of the presence of some of the *Vibrio* sp. in shellfish, no basis for dependence on the indicator standards has been found. In other words, the *Vibrio* sp. have been shown to be natural inhabitants of the brackish water environments near the coast, and these organisms remain there without having to be replenished in fresh discharges of sewage laden waters. For this reason, there has been renewed concern about the safety of consuming raw seafoods, regardless of the site of harvest. As reported by the *FDA Consumer,* the shellfish industry, food industries, and government agencies have expended great effort and much money in attempting to maintain good, safe quality of seafoods. However, the survival of these organisms in otherwise unpolluted coastal waters, the inability of regulatory agencies to completely stop unlawful harvest, and the potential for contamination of shellfish after harvest, as well as the multiplication of potential pathogens in shellfish being shipped or stored, has made it impossible to ensure a completely safe product in all areas at all times.

Freshly harvested fish, shellfish, or crustacea must be kept alive in clean water, or must be immediately iced (melting ice or brine, at $-2°$ C), and in modern harvesting, some fishing vessels are able to immediately freeze the catch, and to keep it frozen until marketing. Many studies have shown the increases that can be expected in microbial counts when seafoods are kept in melting ice. This temperature will greatly slow the multiplication of bacteria, but will not stop their growth. Other studies have shown that some organisms, in some types of seafoods are able to survive freezing temperatures to $-20°$ C for as long as one month to six weeks. It is apparent, then, that although cooling and freezing are most beneficial to prevent spoilage of seafoods for short periods of time, these processes do not remove any contamination that may have occurred. In preventing obvious spoilage, the food may appear to be satisfactory for consumption, when actually it is still harboring an adequate number of bacteria to initiate an infection in the consumer.

Environmental conditions at the location of seafood harvest have much to do with the potential safety of the food for consumption. Generally, where the water and air temperatures are low, that is, in near-arctic regions, the danger of disease transmission to seafood consumers is relatively low. Where the temperatures of air and water are higher, as in tropical or subtropical areas, the dangers to the consumers are higher. In geographical areas where there is known to be much water pollution, and generally much waterborne intestinal disease such as cholera, salmonellosis, hepatitis, poliomyelitis, and so on, the danger that growing waters are polluted, and the danger of transmission of infection are increased.

Generally, whether seafoods are marketed raw, processed, cooked, canned, frozen, or smoked, good manufacturing practices as discussed in an earlier chapter are essential for a good, palatable, safe product. In the case of seafoods, because the initial raw product is likely to be contaminated with potentially pathogenic organisms, it is essential that cleanliness, disinfection, and general specific routines of handling, processing, storing, and preparing be spelled out and rigidly adhered to in all steps of the operation. If one visits a seafood plant, it is sometimes surprising that the conditions within the plant itself are among the cleanest and most appealing that can be imagined in a food plant. If this is not the case, the odor that will greet you before you reach the plant will probably tell you of the conditions within.

The kinds of specific regulations concerning the marketing of shellfish in foodservice establishments are included in the following excerpts from the *Food Ordinance of the City of Houston, Texas.*

(b) Fresh and frozen shucked shellfish (oysters, clams, or mussels) shall be packed in nonreturnable packages identified with the name and address of the original shell stock processor, shucker-packer, or repacker, and the interstate certification number issued according to law. Shell stock and shucked shellfish shall be kept in the container in which they were received until they are used. Each container of unshucked shell stock (oysters, clams, or mussels) shall be identified by an attached tab that states the name and address of the original shell stock processor, the kind and quantity of shell stock, and an interstate certification number issued by the State or foreign shellfish control agency.

Although the intervals between changes may be relatively long, there are periodic changes in regulations governing seafood and shellfish regulations. The FDA has recently overseen the first revision of the National Shellfish Sanitation Program *Manual of Operations* in over 20 years. The first part, *Sanitation of Shellfish Growing Areas,* was published in June 1986, and the second part, *Sanitation of the Harvesting and Processing of Shellfish,* is still being drafted. When completed, it must be approved by the Interstate Shellfish Sanitation Conference, an organization involving representatives of the shellfish-producing states, the FDA, the shellfish industry, and the National Marine Fisheries Service. The revisions have attempted to clarify and strengthen existing regulations for shipping and storing shellfish, and to create uniformity from state to state. In this, as in any other situation, seafood dealers are subject to inspection at all stages of operation and must comply with local to national regulations specifically detailing conditions and operation procedures for their businesses. Examples are seen in the *Houston Food Ordinance* cited previously, and in the following from the *New York State Sanitary Code:*

(c) Shellfish.

(1) Fresh or frozen oysters, clams, and mussels shall meet the standards of quality established for such foods by applicable Federal and State laws and regulations. If the shellfish source is outside the State of New York, it shall be one which appears on the current Federal Food and Drug Administration "Interstate Certified Shellfish Shippers List".

(2) Shell stock shall be identified with an official tag giving the name and certificate number of the original shell stock shipper and the kind and quantity of shell stock. Fresh and frozen shucked oysters, clams, and mussels, shall be packed in non-returnable containers identified with the name and address of the shell stock shipper, shucker-packer, or repacker and the certificate number of the shipper.

(3) Shucked shellfish shall be kept in the original container until used.

(d) Fresh, frozen, or smoked fish, eel, scallop and edible crustacea including lobster, crab, and shrimp shall meet the standards of quality established for such products by applicable Federal, State and local laws, rules and regulations.

In order to enforce the requirements just quoted, it is frequently necessary that Public Health personnel submit shellfish, other seafoods, and/or seawater to laboratories for testing. If the results of the tests are to be meaningful, standard methods must be used in all test procedures. To this end, the American Public Health Association in 1985 published the fifth edition of *Laboratory Procedures for the Examination of Seawater and Shellfish*. In the preface to that book, it is stated that it is the intent of the publishers that the book provide the analytical procedures needed to evaluate shellfish and the waters in which they are raised. Due to the nature of shellfish, both the water and the meats of the food animals must be tested to assure good quality and the absence of potentially pathogenic microorganisms. This publication includes tests for the presence of both bacteria and viruses in the water and shellfish meats. Both the necessary laboratory and field methods are described in detail in the book.

Good manufacturing practices (GMPs) as described in earlier chapters also apply to the processing and shipping of seafoods. In addition to the general requirements discussed earlier, there are specific requirements in Subpart A for smoked and smoke-flavored fish, and for breaded shrimp. These are found in the provisions of Subpart A at 35 FR 17401 and 35 FR 17840 of November 1970.

Seafoods of all types, harvested from clean waters, handled carefully, and kept at the proper temperatures must be considered to be potentially hazardous foods because of their nature. If all requirements are faithfully followed, and the temperature limits are rigidly observed, then the shipping life, if not the shelf life of the product will be lengthened. Fresh, nutritious seafoods are increasingly important in the healthful diets recommended by more and more health experts. To follow all rules and regulations concerning their growth, harvest, shipping, processing, and preparation will continue to insure a plentiful supply of these delightful foods.

BIBLIOGRAPHY

BALLENTINE, C. 1985. Pollution narrows shellfish harvest. *FDA Consumer*. February: 10–13.

CORWIN, E. 1975. For safer shellfish. *FDA Consumer*. October: 9–11.

GREENBERG, A. E., and HUNT, D. A., eds. 1985. *Laboratory procedures for the examination of seawater and shellfish*. Washington, DC: The American Public Health Association.

INTERNATIONAL COMMITTEE ON MICROBIOLOGICAL SAFETY OF FOODS. 1980. In *Microbial ecology of foods*. vol. 2, pp. 567–605. New York: Academic Press.

LARKIN, E. P., and HUNT, D. A. 1982. Bivalve mollusks: Control of microbiological contaminants. *BioScience* 32:193–197.

U.S. GOVERNMENT PRINTING OFFICE. 1974. *Federal Register*. Part 128. Human Foods; Current Good Manufacturing Practice (Sanitation) in Manufacture, Processing, Packing, or Holding. Department of Health, Education, and Welfare. Food and Drug Administration. Subpart A. Washington, DC.

WOOD, P. D., ed. 1976. *Guide to shellfish hygiene*. Geneva, Switzerland: World Health Organization.

14

Training Programs

With the mushrooming numbers of foodservice establishments, not only in the United States but in many other countries, it becomes more and more imperative that persons who work in this industry have some training in at least the mechanics of food sanitation. It would be preferable if these employees had some good training in the principles of sanitation; however, with the low pay and fast turnover in personnel in many food establishments, it is unlikely that we will ever reach that goal. It is possible, though, for managers of foodservice establishments to receive some training, and learn some local ordinances. Houston, Texas, for example, now requires that managers of foodservice businesses undergo a training course in the rudiments of food sanitation practices. Retail foodservice businesses become more and more competitive in times of economic slowdown, and certainly those whose managers can take advantage of all positive attributes for a business are going to be much better off and are likely to remain in competition for longer periods of time.

By the mid-1970s the foodservice industry was the largest retail industry in terms of numbers of persons employed, with approximately four million employees serving an average of 150 million meals each day in the United States. Because of the varied types of foodservice operations, it is apparent that many of the four million workers were employed with minimum training, even in the mechanics of their jobs, but with time have become adept at such mechanics. Because no formal training or education is usually required, a job in the foodservice industry is frequently the first employment for young people, and often on a part-time basis while these persons are still attending high school. The fact that these young people do an adequate job is readily observed in the numbers of them who continue to be hired, and rehired in this industry. On the other hand, the age, and the multiple responsibilities of school, home, and work lead to a very rapid employee turnover and job change rate in this industry.

Also involved in this employee turnover rate is the lack of required training and education, with therefore, an understandably lower pay scale. The combination of low pay and the temporary nature of employment in much of the industry, added to sometimes inconvenient work hours (i.e., working on holidays when friends are free, possible split-shift work, late hours, etc.) make the high turnover rate easily understandable. As a result, there is frequent need to train new employees in the mechanics of the job, and this type of training becomes an ongoing program in many foodservice establishments. Training in principles of sanitation, which is badly needed, is all too often never begun.

The Food and Drug Administration recommended in the *1976 Food Service Sanitation Manual* that the foodservice industry, together with federal, state, and local government agencies, plan effective programs to give basic training in principles of sanitation and provide continued in-service training in putting these principles into practice. In 1976 the FDA awarded a contract to the National Institute for the Foodservice Industry (NIFI) to develop a plan for implementing a uniform national program of sanitation training and certification of owners, operators, and managers of foodservice establishments.

The NIFI developed such a plan and it was submitted as a final report to the FDA in 1977. In addition to that program, the Educational Institute of the American Hotel and Motel Association and the Foodservice Certification Society associated with the Oklahoma Restaurant Association also offered programs, as do a number of health departments, universities, and foodservice companies. The City of Houston, Texas has developed, and since 1978 has required participation in such a program for foodservice managers. Through December 1976, over 68,000 persons had completed some type of training and had been certified through these programs, and many thousands more have been added to the certified list since that time. These numbers still represent only a small percentage of the four million workers, and too few cities and states require certification of any workers.

Although the ideal would be to have mandatory certification for all foodservice workers, the owners are perhaps justified at this time in objecting that the cost would be too great when the benefits of training all workers are considered. Certification of managers, however, is a step toward accomplishing the same objective, and if the managers are then required to supervise in-service training for other workers, the objective of preventing, or at least reducing the transmission of foodborne diseases can be met. Such programs would reduce potential costs of training, but properly done will provide benefits to the owner in food savings and customer satisfaction, and to the customer in better food with reduced hazard of foodborne disease.

As the foodservice industry grows and the needs for sanitation programs are more widely realized, it is to be hoped that the numbers of local and state governments with mandatory requirements will increase. Persons concerned with the health of consumers continue to urge training for certification, at least for managers, as one important means of controlling foodborne disease transmission. During periods of economic slumps, with business more competitive, and volume of business reduced, training programs are among the first cost-cutting measures that may be dropped. Such operations, however, may be false economy in that there will be increased danger of foodborne disease outbreak.

Several suitable training course outlines are available. A good course can be produced by using the recommendations of the FDA in the *1976 Foodservice Sanitation Manual*. This manual was written in sufficient detail to give the managers (a) the model governmental ordinance, (b) what each recommended section will accomplish when properly complied with, and (c) why this result is desirable. The City of Houston, Texas has developed a good one-day program for training and certifying foodservice managers. That program is required for foodservice managers within the city, and a fee is charged for the program in order to help support it.

Regardless of which program is used, an effective program will contain elements to teach certain basic elements of sanitation, and food handling. The following contains the major components of an effective training program:

SUGGESTED TOPICS AND TIME OF TRAINING FOR A FOODSERVICE MANAGER TRAINING PROGRAM

I. Foodborne Disease: What causes it; what can prevent it?
   A. Food poisoning—bacterial
   B. Food poisoning—chemical
   C. Foodborne infection   2 hours

II. Food Protection: How and why?
   A. Potentially hazardous foods: Why?
   B. Contamination of food: Where and when it happens; relationship of when it happens to probable result.   2 hours

III. Food Storage.
   A. Dry, canned, bottled.
   B. Refrigerated.
   C. Frozen.
   D. Heated.   1 hour

IV. Food Preparation.
   A. Handling, raw, dry, liquids, frozen, potentially haz-
      ardous.
   B. Cooking, thawing, storing, reheating.   1 hour

V. Food Display and Service.
   A. Temperatures, thermometers.
   B. Dispensing, milk, non-dairy, condiments, ice, utensils.
   C. Re-service.   1 hour

VI. Food Transport.   1 hour

VII. Personnel.
   A. Health.
   B. Personal cleanliness.
   C. Clothing and appearance.
   D. Employee practices.
   E. In-service training.   2 hours

VIII. Equipment and Utensils.
   A. Materials, single-service items.
   B. Design, fixed, movable.
   C. Ventilation, cleaning schedule.
   D. Refrigeration, cleaning schedule.
   E. Installation and location.   1 hour

IX. Cleaning, Sanitizing Equipment and Utensils.
   A. Each-use cleaning and sanitizing.
   B. Periodic cleaning and sanitizing.
   C. Cleaning cloths, manual cleaning.
   D. Mechanical cleaning and sanitizing.
   E. Drying.   1 hour

X. Storing Equipment and Utensils.
   A. Reuse items.
   B. Single-service items.
   C. Prohibited storage.   1 hour

XI. Sanitary Facilities.
   A. Water supply. Why?
   B. Sewage, plumbing.

C. Toilet facilities.
D. Lavatory facilities.
E. Garbage and refuse.
F. Insect and rodent control.   1 hour

XII. Physical Facilities. Design and Cleaning.
A. Floors, drains.
B. Walls, ceilings.
C. Prohibited items.
D. Lighting.
E. Ventilation.
F. Dressing rooms.   1 hour

XIII. Poisonous and Toxic Materials.
A. Prohibited.
B. Labeling.
C. Storage.   1 hour

XIV. Compliance, Inspections, Management, Testing.
A. Inspection schedules.
B. Inspection reports.
C. Correcting deficiencies.
D. Condemning food.
E. Economics of compliance and training.
F. Evaluation and testing.   2 hours

   TOTAL TIME   18 HOURS

The completion of a training course containing these subjects, while it will not make anyone expert, will improve the sanitary quality of a foodservice operation if the manager will see that the principles are applied. There are portions of the course that many managers may not have an opportunity to apply, such as design and location of equipment, and sanitary facilities, and design of the physical plant. The manager, however, may well get some innovative ideas as to how an already established operation may be improved by rearrangement and/or initiation of sanitary practices to fit the surroundings. Once having completed the course and having been certified, the manager should now return to the place of business and step by step, apply the newly gained knowledge to this specific operation. Having done this, the stage is now set, and all necessary props are at hand for an innovative manager to take any untrained employees through a much shortened, but effective in-service program to improve and maintain a desirable level of sanitation, and full compliance with the letter and spirit of governmental regulations concerning the operation of a foodservice business.

BIBLIOGRAPHY

NATIONAL INSTITUTE FOR THE FOODSERVICE INDUS-
TRY. 1977. *Development of a uniform national plan for sanitation
training of foodservice managers.* Final Report for FDA Contract 223-76-
2072. Chicago: Author.

TEXAS RESTAURANT ASSOCIATION. 1978. *Food service sanitation
manual.* College Station, TX: Texas A&M University.

THE CITY OF HOUSTON, TEXAS. 1978. *Food service manager's certification manual.* City of Houston Health Department.

U.S. DEPARTMENT OF HEALTH, EDUCATION, AND WEL-
FARE. 1976. *Food service sanitation manual.* Public Health Service.
Food and Drug Administration. DHEW Publication no. (FDA) 78-2081.
Washington, DC: U.S. Government Printing Office.

Appendix I

Glossary

Adulterate   To corrupt, debase, or make impure by addition of a foreign substance.

Antibiotic   A chemical compound produced by a living organism that interferes with growth or life of another living organism. Used to treat some infectious diseases.

Antiseptic   A chemical substance used to interfere with, or to inhibit the growth of some microorganisms. May be used on some living tissues, as in the human body.

Bactericide   A chemical that will kill some bacterial cells.

Carrier   An apparently healthy human who harbors, in or on his or her body, pathogenic microorganisms, and who may spread those pathogens to susceptible humans.

Chronic disease   A disease that progresses extremely slowly if at all.

Contaminate   To add foreign and unwanted matter to any object or environment.

Detergent   A chemical cleanser similar to soaps, but of different chemical nature.

Disease   A condition in which bodily health is impaired; sickness, illness.

Disinfect   To remove potentially pathogenic organisms from an object, or an environment.

Disinfectant   A chemical used to remove or kill potentially pathogenic organisms from objects or environment.

Epidemic   A larger than usual number of cases of a particular infectious disease in one locality.

Food   Any raw, cooked, or processed edible substance, ice, beverage, or ingredient used or intended for use or for sale in whole or in part for human consumption.

Fungicide   A chemical that will kill some kinds of fungal cells.

Germicide   A chemical that will kill some kinds of microorganism cells.

GRAS substances   Food additives, generally regarded as safe for use.

Immunization   The deliberate exposure of a living animal to disease-producing agents in a harmless state, to permit the animal to develop immunity or resistance to that agent.

Infection   The condition produced when microorganisms invade tissues and multiply within those tissues.

Infestation   The presence and multiplication of unwanted living organisms in any location.

Insecticide   A chemical used to kill insects.

Normal flora   Those microorganisms that are usually present in an environment, or in or on the human body, under "normal" or non-disease circumstances.

Opportunist   A microorganism that exists frequently, or continuously, in or on the human body in a harmless state that produces disease when provided the opportunity by altered conditions in its environment.

Pasteurization   Heating fluids to a temperature that will kill most pathogenic organisms.

Pathogen   A microorganism that is capable of producing disease when it enters the human or animal body.

Pesticide   A chemical that will kill some kind of living organism constituting a nuisance to the human.

Pollution   The accumulation of any foreign, unwanted matter in an environment in which it becomes a nuisance or a danger to the health of the environment.

Potentially hazardous food   Any food that consists in whole or in part of milk or milk products, eggs, meat, poultry, fish, shellfish, edible crustacea, or other ingredients, including synthetic ingredients, in a form capable of supporting rapid and progressive growth of infectious or toxigenic microorganisms. The term does not include clean, whole, uncracked, odor-free shell eggs or foods that have a pH of 4.6 or below or a water activity (a_w) value of 0.85 or less.

Preservative   A chemical additive that will preserve some desired characteristic of a food by making certain chemical structures more stable or by retarding microbial growth.

Rodenticide   A chemical that will kill rodents.

Sanitary   Harmless to human health and well-being, and of good quality.

Sanitation   The use of practices that will make an environment or substance harmless to human health and well-being.

Sanitize   Treatment by heat or chemicals for a long enough time to reduce the number of bacteria present.

Saprophyte   An organism that obtains its food from nonliving matter.

Sterilize   To remove all forms of life from an environment, or to kill all forms of life in an environment.

Toxin   A chemical produced by living organisms that is poisonous to man or animals.

Appendix II

Common Infectious Diseases That May Be Foodborne

Appendix II Common Infectious Diseases That May Be Foodborne

Disease	Synonym	Causative Agent	Mode of Transmission	Methods of Prevention	Treatments Available
Usually Foodborne					
Food poisoning	Staph food poisoning	Soluble enterotoxin produced by growth of *Staphylococcus aureus* in foods	Ingestion of contaminated foods	1. Prevent contamination 2. Refrigerate	Supportive
Food poisoning		Growth of *C. perfringens* in foods, most often meats	Ingestion of contaminated foods	1. Prevent contamination 2. Serve foods hot without delay 3. Cook adequately 4. Refrigerate	Supportive
Botulism		Soluble toxins produced by growth of *C. botulinum* in anaerobic nonacid foods	Ingestion of contaminated foods	1. Prevent contamination 2. Proper heat preservation 3. Heat to boiling for 15 min. before eating	Specific antitoxin
Bacillus cereus		Cell associated toxins in intestine	Ingestion of contaminated foods	1. Prevent contamination 2. Refrigerate	Supportive
Cholera	Gastroenteritis	*Vibrio cholerae*	Ingestion of contaminated food or water	1. Properly treat water 2. Properly cook food	Antibiotic and supportive
Vibrio parahaemolyticus food poisoning		Cell-associated toxicity	Ingestion of contaminated foods	1. Adequately cook seafoods 2. Refrigerate seafoods until cooked	Antibiotic and supportive

Appendix II *(continued)*

Disease	Synonym	Causative Agent	Mode of Transmission	Methods of Prevention	Treatments Available
Salmonellosis	Food poisoning	Any one of many species or types of *Salmonella*	Ingestion of live organisms in contaminated foods or water	1. Prevent contamination 2. Clean raw foods 3. Cook thoroughly 4. Refrigerate 5. Detect and eliminate carriers	Antibiotic treatment has irregular success
Typhoid fever, Paratyphoid fever	Enteric fever	*Salmonella typhi, Salmonella paratyphi A, Salmonella paratyphi B, Salmonella paratyphi C*	Ingestion of live organisms in contaminated foods or water	1. Chlorinate water 2. Detect and eliminate carriers 3. Cook foods properly 4. Immunize 5. General sanitation	Antibiotic treatment
Campylobacteriosis	Gastroenteritis	*Campylobacter jejuni*	Frequently raw milk	1. Consume only pasteurized milk 2. Cook all foods properly	Antibiotic
Shigellosis	Bacillary dysentery	Any one of many species or types of *Shigella*	Ingestion of live organisms in contaminated foods or water	1. Chlorinate water 2. Cook and handle foods properly 3. General sanitation	Antibiotic treatment plus fluid maintenance
Streptococcal pharangitis	Strep throat, Septic sore throat	*Streptococcus pyogenes*, many types	Ingestion of live organisms in contaminated food or milk. Also contact and respiratory	1. General sanitation 2. Pasteurize milk 3. Cook, handle, and store food properly	Antibiotic treatment

Disease	Organism	Common name	Mode of transmission	Control	Treatment
Diphtheria	*Corynebacterium diphtheriae*		Ingestion of live organisms in milk or food. Also contact and respiratory	1. Immunize 2. Pasteurize milk 3. Handle and refrigerate food properly 4. General sanitation	Antibiotic treatment and antitoxin treatment
Brucellosis	*Brucella abortus B. melitensis or B. suis*	Undulent fever, milk fever, Malta fever	Ingestion of live organisms in milk or meat products. Also contact	1. Pasteurize milk 2. Cook milk and meat products properly 3. General sanitation	Antibiotic treatment
Infectious hepatitis	Virus	Epidemic jaundice, Catarrhal jaundice, Hepatitis A	Ingestion of virus in contaminated water, milk and food. Also direct contact	1. General sanitation 2. Isolate cases	Gamma globulin
Amoebiasis	*Entamoeba histolytica*	Amoebic dysentery	Ingested cysts of organism in contaminated water, food	1. General sanitation 2. Water filtration	Antibiotic and chemical therapy
Acute diarrheal disease	*Escherichia coli, Shigella sp., Salmonella sp., Giardia lamblia, Staphylococcus sp., Pseudomonas aeruginosa, Proteus vulgaris, Vibrio sp., others*	Summer complaint, Travelers' diarrhea, Infant diarrhea	Ingestion of live organisms in contaminated water, food. Also direct contact	1. Chlorinate water 2. General sanitation	Antibiotic and supportive therapy, especially fluid balance in children

Appendix II (*continued*)

Disease	Synonym	Causative Agent	Mode of Transmission	Methods of Prevention	Treatments Available
Trichinosis	Trichiniasis, Trichinellosis	Larva of *Trichinella spiralis*	Ingested meat containing viable larva of organisms	1. Process pork adequately 2. Cook pork adequately	Antibiotic treatment
Epidemic gastroenteritis	The virus	One of several viruses	Ingestion of virus in contaminated water, food. Contact.	1. Treat water 2. General sanitation	Supportive therapy
Sometimes Foodborne					
Tularemia		*Francisella tularensis*	Direct contact Insect bite Ingestion of organisms in meat of infected animal Contaminated water	1. Cook meat, especially rabbit, properly 2. Chlorinate water	Antibiotic therapy
Tuberculosis	TB	*Mycobacterium tuberculosis*	Contact Respiratory Consumption of organism in milk from infected cows	1. General sanitation 2. Pasteurize milk and milk products 3. Eliminate infected cattle	Antibiotic and supportive therapy
Scarlet fever		*Streptococcus* sp.	Direct contact Droplet inhalation Airborne	1. General sanitation	Antibiotic treatment

Disease	Common name	Causative agent	Transmission	Prevention/Control	Treatment
Strep throat	Sore throat	*Streptococcus* sp., other bacteria	Direct contact Droplet inhalation Airborne Indirect through contaminated articles, contaminated milk or food	1. General sanitation	Antibiotic treatment
Common cold		Any one of many different viruses	Direct contact Droplet inhalation Indirect through contaminated articles including eating utensils	1. General sanitation	None
Influenza	Flu	One of several types of influenza virus	Direct contact Droplet inhalation Airborne Indirect through contaminated articles including eating utensils	1. General sanitation 2. Immunization	None
Moniliasis	Candidiasis, Thrush	*Candida albicans*	Direct contact Indirectly through contaminated articles	1. General sanitation	Antibiotic chemical treatment

291

Appendix III

Selected Representative Laws, Codes, and Ordinances Governing Food Industries

U.S. FOOD, DRUG, AND COSMETIC ACT, AS AMENDED

Chapter III—Prohibited Acts and Penalties

Prohibited Acts

Sec. 301 [331]. The following acts and the causing thereof are hereby prohibited:

(a) The introduction or delivery for introduction into interstate commerce of any food, drug, device, or cosmetic that is adulterated or misbranded.

(b) The adulteration or misbranding of any food, drug, device, or cosmetic in interstate commerce.

(c) The receipt in interstate commerce of any food, drug, device, or cosmetic that is adulterated or misbranded, and the delivery or proffered delivery thereof for pay or otherwise.

Chapter IV—Food

Definitions and Standards for Food

Sec. 401 [341]. Whenever in the judgment of the Secretary such action will promote honesty and fair dealing in the interest of consumers, he shall promulgate regulations[3] fixing and establishing for any food, under its common or usual name so far as practicable, a reasonable definition and standard of identity, a reasonable standard of quality, and/ or reasonable standards of fill container: *Provided,* That no definition and standard of identity and no standard of quality shall be established for fresh or dried fruits, fresh or dried vegetables, or butter, except that definitions and standards of identity may be established for avocados, cantaloupes, citrus fruits, and melons. In prescribing any standard of fill of container, the Secretary shall give due consideration to the natural shrinkage in storage and in transit of fresh natural food and to need for the necessary packing and protective material. In the prescribing of any standard of quality for any canned fruit or canned vegetable, consideration shall be given and due allowance made for the differing characteristics of the several varieties of such fruit or vegetable. In prescribing a definition and standard of identity for any food or class of food in which optional ingredients are permitted, the Secretary shall, for the purpose of promoting honesty and fair dealing in the interest of consumers, designate the optional ingredients which shall be named on the label. Any definition and standard of identity prescribed by the Secretary for avocados, cantaloupes, citrus fruits, or melons shall relate only to maturity and to the effects of freezing.

Adulterated food

Sec. 402 [342]. A food shall be deemed to be adulterated—

(a) (1) If it bears or contains any poisonous or deleterious substance which may render it injurious to health; but in case the substance is not an added substance such food shall not be considered adulterated under this clause if the quantity of such substance in such food does not ordinarily render it injurious to health; or

(2) (A) if it bears or contains any added poisonous or added deleterious substance (other than one which is (i) a pesticide chemical in or on a raw agricultural commodity; (ii) a food additive; or (iii) a color additive) which is unsafe within the meaning of section 406, or (B) if it is a raw agricultural commodity and it bears or contains a pesticide chemical which is unsafe within the meaning of section 408(a); or (C) if it is, or it bears or contains, any food additive which is unsafe within the meaning of section 409; *Provided,* That where a pesticide chemical has been used in or on a raw agricultural commodity in conformity with an exemption granted or a tolerance prescribed under section 408 and such raw agricultural commodity has been subjected to processing such as canning, cooking, freezing, dehydrating, or milling, the residue of such pesticide chemical remaining in or on such processed food shall, notwithstanding the provisions of sections 406 and 409, not be deemed unsafe if such residue in or on the raw agricultural commodity has been removed to the extent possible in good manufacturing practice and the concentration of such residue in the processed food when ready

to eat is not greater than the tolerance prescribed for the raw agricultural commodity; or

(3) if it consists in whole or in part of any filthy, putrid, or decomposed substance, or if it is otherwise unfit for food; or (4) if it has been prepared, packed, or held under insanitary conditions whereby it may have become contaminated with filth, or whereby it may have been rendered injurious to health; or (5) if it is, in whole or in part, the product of a diseased animal or of an animal which has died otherwise than by slaughter; or (6) if its container is composed, in whole or in part, of any poisonous or deleterious substance which may render the contents injurious to health; or (7) if it has been intentionally subjected to radiation, unless the use of the radiation was in conformity with a regulation or exemption in effect pursuant to section 409.[4]

(b) (1) If any valuable constituent has been in whole or in part omitted or abstracted therefrom; or (2) if any substance has been substituted wholly or in part therefor; or (3) if damage or inferiority has been concealed in any manner; or (4) if any substance has been added thereto or mixed or packed therewith so as to increase its bulk or weight, or reduce its quality or strength, or make it appear better or of greater value than it is.

(c) If it is, or it bears or contains, a color additive which is unsafe within the meaning of section 706(a).

(d) If it is confectionery, and—

(1) has partially or completely imbedded therein any nonnutritive object: *Provided,* That this clause shall not apply in the case of any nonnutritive object if, in the judgment of the Secretary as provided by regulations, such object is of practical functional value to the confectionery produce and would not render the product injurious or hazardous to health;

(2) bears or contains any alcohol other than alcohol not in excess of one-half of 1 per centum by volume derived solely from the use of flavoring extracts; or

(3) bears or contains any nonnutritive substance: *Provided,* That this clause shall not apply to a safe nonnutritive substance which is in or on confectionery by reason of its use for some practical functional purpose in the manufacture, packaging, or storage of such confectionery if the use of the substance does not promote deception of the consumer or otherwise result in adulteration or misbranding in violation of any provision of this Act: *And provided further,* That the Secretary may, for the purpose of avoiding or resolving uncertainty as to the application of this clause, issue regulations allowing or prohibiting the use of particular nonnutritive substances.

(e) If it is oleomargarine or margarine or butter any of the raw material used therein consisted in whole or in part of any filthy, putrid, or decomposed substance, or such oleomargarine or margarine or butter is otherwise unfit for food.

Tolerances for Poisonous Ingredients in Food

Sec. 406 [346]. Any poisonous or deleterious substance added to any food, except where such substance is required in the production thereof or cannot be avoided by good manufacturing practice shall be deemed to be unsafe for purposes of the application of clause (2) (A) of section 402(a); but when such substance is so required or cannot be so avoided, the Secretary shall promulgate regulations[7] limiting the quantity therein or thereon to such extent as he finds necessary for the protection of public health, and any quantity exceeding the limits so fixed shall also be deemed to be unsafe for purposes of the application of clause (2) (A) of section 402(a). While such a regulation is in effect limiting the quantity of any such substance in the case of any food, such food shall not, by reason of bearing or containing any added amount of such substance, be considered to be adulterated within the meaning of clause (1) of section 402(a). In determining the quantity of such added substance to be tolerated in or on different articles of food the Secretary shall take into account the extent to which the use of such substance is required or cannot be avoided in the production of each such article, and the other ways in which the consumer may be affected by the same or other poisonous or deleterious substances.

Tolerances for Pesticide Chemicals In or On Raw Agricultural Commodities

Sec. 408 [346a]. (a) Any poisonous or deleterious pesticide chemical, or any pesticide chemical which is not generally recognized, among experts qualified by scientific training and experience to evaluate the safety of pesticide chemicals, as safe for use, added to a raw agricultural commodity, shall be deemed unsafe for the purposes of the application of clause (2) of section 402(a) unless—

(1)   a tolerance for such pesticide chemical in or on the raw agricultural commodity has been prescribed by the Secretary of Health, Education, and Welfare under this section[8] and the quantity of such pesticide chemical in or on the raw agricultural commodity is within the limits of the tolerance so prescribed; or

(2)   with respect to use in or on such raw agricultural commodity, the pesticide chemical has been exempted from the requirement of a tolerance by the Secretary under this section.

While a tolerance or exemption from tolerance is in effect for a pesticide chemical with respect to any raw agricultural commodity, such raw agricultural commodity shall not, by reason of bearing or containing any added amount of such pesticide chemical, be considered to be adulterated within the meaning of clause (1) of section 402(a).

(b) The Secretary shall promulgate regulations[8] establishing tolerances with respect to the use in or on raw agricultural commodities of poisonous or deleterious pesticide chemicals and of pesticide chemicals which are not generally recognized, among experts qualified by scientific training and experience to evaluate the safety of pesticide chemicals, as safe for use, to the extent necessary to protect the public health. In establishing any such regulation, the Secretary shall give appropriate consideration, among other relevant factors, (1) to the necessity for the production of an adequate, wholesome, and economical food supply; (2) to the other ways in which the consumer may be affected by the same pesticide chemical or by other related substances that are poisonous or deleterious; and (3) to the opinion of the Secretary of Agriculture as submitted with a certification of usefulness under subsection (1) of this section. Such regulations shall be promulgated in the manner prescribed in subsection (d) or (e) of this section. In carrying out the provisions of this section relating to the establishment of tolerances, the Secretary may establish the tolerance applicable with respect to the use of any pesticide chemical in or on any raw agricultural commodity at zero level if the scientific data before the Secretary does not justify the establishment of a greater tolerance.

(c) The Secretary shall promulgate regulations exempting any pesticide chemical from the necessity of a tolerance with respect to use in or on any or all raw agricultural commodities when such a tolerance is not necessary to protect the public health. Such regulations shall be promulgated in the manner prescribed in subsection (d) or (e) of this section.

Other specifications in addition to the above follow this article in the complete Act.

TEXAS FOOD, DRUG, AND COSMETIC ACT

Prohibited Acts

Sec. 3. The following acts and the causing thereof within the State of Texas are hereby declared unlawful and prohibited:

(a) The manufacture, sale, or delivery, holding or offering for sale of any food, drug, device, or cosmetic that is adulterated or misbranded;

(b) The adulteration or misbranding of any food, drug, device, or cosmetic;

(c) The receipt in commerce of any food, drug, device, or cosmetic that is adulterated or misbranded, and the delivery or proffered delivery thereof for pay or otherwise;

Food—Adulteration Defined

Sec. 10. A food shall be deemed to be adulterated:

(a) (1) If it bears or contains any poisonous or deleterious substance which may render it injurious to health; but in case the substance is not an added substance such food shall not be considered adulterated under this clause if the quantity of such substance in such food does not ordinarily render it injurious to health; or (2) (A) if it bears or contains any added poisonous or added deleterious substance (except a pesticide chemical in or on a raw agricultural commodity and except a food additive) which is unsafe within the meaning of Section 13; or (B) if it is a raw agricultural commodity and it bears or contains a pesticide chemical which is unsafe within the meaning of Section 13, or (C) if it is, or it bears or contains, any food additive which is unsafe within the meaning of Section 13; provided, that where a pesticide chemical has been used in or on a raw agricultural commodity in conformity with an exemption granted or a tolerance prescribed under Section 13, and such raw agricultural commodity has been subjected to processing such as canning, cooking, freezing, dehydrating, or milling the residue of such pesticide chemical remaining in or on such processed food shall, notwithstanding the provisions of Section 13, and clause (C) of this Section, not be deemed unsafe if such residue in or on the raw agricultural commodity has been removed to the extent possible in good manufacturing practice, and the concentration of such residue in the processed food, when ready to eat, is not greater than the tolerance prescribed for a raw agricultural commodity; or (3) if it consists in whole or in part of a diseased, contaminated, filthy, putrid, or decomposed substance, or if it is otherwise unfit for foods; or (4) if it has been produced, prepared, packed or held under unsanitary conditions whereby it may have become contaminated, or whereby it may have been rendered injurious to health; or (5) if it is the product of a diseased animal or an animal which has died otherwise than by slaughter, or that has been fed upon the uncooked offal from a slaughterhouse; or (6) if its container is composed, in whole or in part, of any poisonous or deleterious substance which may render the contents injurious to health.

(b) (1) If any valuable constituent has been in whole or in part omitted or abstracted therefrom; or (2) if any substance has been substituted wholly or in part therefor; or (3) if damage or inferiority has been concealed in any manner; or (4) if any substance has been added thereto or mixed or packed therewith so as to increase its bulk or weight, or reduce its quality or strength or make it appear better or of greater value than it is; or (5) if it contains saccharin, dulcin, glucin, or other sugar substitutes except in dietary foods, and when so used shall be declared; or (6) if it be fresh meat and it contains any chemical substance containing sulphites, sulphur dioxide, or any other chemical preservative which is not approved by the United States Bureau of Animal Industry or the Commissioner of Health.

(c) If it is confectionery and it bears or contains any alcohol or nonnutritive article or substance except harmless coloring, harmless flavoring, harmless resinous glaze not in excess of four-tenths of one per centum, harmless natural gum, and pectin; provided, that this paragraph shall not apply to any confectionery by reason of its containing less than one-half of one per centum by volume of alcohol derived solely from the use of flavoring extracts, or to any chewing gum by reason of its containing harmless nonnutritive masticatory substances.

(d) If it bears or contains a coal-tar color other than one certified under authority of the Federal Act.

STATE OF GEORGIA RULES OF DEPARTMENT OF HUMAN RESOURCES—PHYSICAL HEALTH

Chapter 290-5-14 Food Service

Note: Any reference in these rules and regulations to the Department of Public Health and the Board of Health means the Department of Human Resources and the Board of Human Resources in accordance with Georgia Laws 1972, pp. 1015, 1032 (Chapter 1, Section 32, Executive Reorganization Act), and with Georgia Laws 1972, pp. 1069, 1071, 1072 (Sections 3 and 4).

290-5-14-.01   Definitions. Amended.   The following definitions shall apply in the interpretation and enforcement of this rule:

(1) "Adulterated" means the condition of a food:

(a) if it bears or contains any poisonous or deleterious substance in a quantity which may render it injurious to health;

(b) if it bears or contains any **added** poisonous or deleterious substance for which no safe tolerance has been established by regulations, or in excess of such tolerance if one has been established;

(c) if it consists in whole or in part of any filthy, putrid, or decomposed substance, or if it is otherwise unfit for human consumption;

(d) if it has been processed, prepared, packed, or held under insanitary conditions, whereby it may have become contaminated with filth, or whereby it may have been rendered injurious to health;

(e) if it is in whole or in part the product of a diseased animal, or an animal which has died otherwise than by slaughter;

(f) if its container is composed in whole or in part of any poisonous or deleterious substance which may render the contents injurious to health.

(2) "Approved" means acceptable to the health authority based on its determination as to conformance with appropriate standards and good public health practice.

(3) "Closed" means fitted together snugly leaving no openings large enough to permit the entrance of flies, roaches, rodents, or other vermin.

(4) "Corrosion-Resistant Material" means a material which maintains its original surface characteristics under prolonged influence of the food, cleaning compounds, and sanitizing solutions which may contact it.

(5) "County Board of Health" means the Board of Health as established by 88-201 and 88-202, Georgia Health Code (Acts. 1964, pp. 499, 512).

(6) "Department" means the Department of Public Health, State of Georgia.

(7) "Easily Cleanable" means readily accessible and of such material and finish, and so fabricated that residue may be completely removed by normal cleaning methods.

(8) "Employee" means any person working in a food-service establishment who transports food or food containers, who engages in food preparation or service, or who comes in contact with any food utensils or equipment.

(9) "Equipment" means all stoves, ranges, hoods, meatblocks, tables, counters, refrigerators, sinks, dishwashing machines, steam tables, and similar items, other than utensils, used in the operation of a food-service establishment.

(10) "Food" means any raw, cooked, or processed edible substances, beverage or ingredient used or intended for use or for sale in whole or in part for human consumption.

(11) "Food-Contact Surfaces" means those surfaces of equipment and utensils with which food normally comes in contact, and those surfaces with which food may come in contact and drain back onto surfaces normally in contact with food.

(12) "Food-Processing Establishment" means an establishment in which food is processed or otherwise prepared and packaged for human consumption.

(13) "Food-Service Establishment" means and includes establishments for the preparation and serving of meals, lunches, short orders, sandwiches, frozen desserts, or other edible products. The term includes but is not limited to restaurants; coffee shops; cafeterias; short order cafes; luncheonettes; taverns; lunch rooms, places manufacturing wholesaling, or retailing sandwiches or salads; soda fountains; institutions, both public and private; food carts; itinerant restaurants; industrial cafeterias; catering establishments; food vending machines and vehicles and operations connected therewith, and similar facilities by whatever name called.

(14) "Health Authority" means the Department, or the County Board of Health acting as its agent.

(15) "Kitchenware" means all multiuse utensils other than tableware used in the storage, preparation, conveying, or serving of food.

(16) "Misbranded" means the presence of any written, printed, or graphic matter, upon or accompanying food or containers of food, which is false or misleading, or which violates any applicable State or local labeling requirements.

(17) "Perishable Food" means any food of such type or in such condition as may spoil.

(18) "Permit" means authorization granted by the health authority to the management to operate a food service establishment and signifies satisfactory compliance with these rules and regulations.

(19) "Person" means any individual, firm, partnership, corporation, trustee or association, or combination thereof.

(20) "Potentially Hazardous Food" means any perishable food which consists in whole or in part of milk or milk products, eggs, meat, poultry, fish, shellfish, or other ingredients capable of supporting rapid and progressive growth of infectious or toxigenic microorganisms.

(21) "Safe Temperatures," as applied to potentially hazardous food, means temperatures of 45°F. or below, and 140°F. or above.

(22) "Sanitize" means effective bactericidal treatment of clean surfaces of equipment and utensils by a process which has been approved by the health authority as being effective in destroying microorganisms, including pathogens.

(23) "Sealed" means free of cracks or other openings which permit the entry or passage of moisture.

(24) "Single-Service Articles" means cups, containers, lids or closures; plates, knives, forks, spoons, stirrers, paddles, straws, placemats, napkins, doilies, wrapping materials; and all similar articles which are constructed wholly or in part from paper, paperboard, molded pulp, foil, wood, plastic, synthetic, or other readily destructible materials, and which are intended by the manufacturers and generally recognized by the public as for one usage only, then to be discarded.

(25) "Tableware" means all multiuse eating and drinking utensils, including flatware (knives, forks and spoons).

(26) "Temporary Food-Service Establishment" means any food-serve establishment which operates at a fixed location for a temporary period of time, not to exceed two (2) weeks, in connection with a fair, carnival, circus, public exhibition, or similar transitory gathering.

(27) "Utensil" means any tableware and kitchenware used in the storage, preparation, conveying, or serving food.

(28) "Wholesome" means in sound condition, clean, free from adulteration, and otherwise suitable for use as human food.

Authority Ga. L. 1964, pp. 507, 559. **Administrative History.** Original Rule entitled "Purpose" was filed and effective on July 19, 1965 as 270-5-6-.01. **Amended:** Rule repealed and a new Rule entitled "Definitions" adopted. Filed January 24, 1967; effective February 12, 1967. **Amended:** Rule renumbered as 290-5-14-.01. Filed June 10, 1980; effective June 30, 1980.

290-5-14-.02 Provisions. Amended.

(1) Permit:

(a) the food-service establishment shall be in satisfactory compliance with the provisions of these rules and the provisions at law which apply to the location, construction and maintenance of food service establishments and the safety of persons therein;

(b) prior to the issuance of a permit and at the request of the health authority the management of the food service establishment shall furnish to the health authority evidence of satisfactory compliance with any law or regulation thereunder applicable to food service establishments but the enforcement of which is the responsibility of a department or agency of government other than the health authority;

(c) the permit shall be framed and publicly displayed at all times;

(d) permits are not transferable from one person to another, from one food service establishment to another nor valid when the food service establishment is moved from one location to another;

(e) the permit shall be returned to the health department when the food service establishment ceases to operate or is moved to another location.

(2) Application for a permit:

(a) the management of the food service establishment shall submit to the health authority an application for a permit;

(b) the application shall be prepared in duplicate on forms provided by the Department. The original shall be forwarded to the health authority and the carbon retained by the management;

(c) the application shall be submitted to the health authority at least ten (10) days prior to the anticipated date of opening and commencement of the operation of the food service establishment;

(d) the application for a temporary food service establishment shall show the inclusive dates of the proposed operation.

Authority Ga. L. 1964, pp. 507, 559. **Administrative History.** Original Rule entitled "Policy" was filed and effective on July 19, 1965 as 270-5-6-.02. **Amended:** Rule repealed and a new Rule entitled "Provisions" adopted. Filed January 24, 1967; effective February 12, 1967. **Amended:** Rule renumbered as 290-5-14-.02. Filed June 10, 1980; effective June 30, 1980.

290-5-14-.03 Food Supplies. Amended.

(1) General:

(a) food in the food-service establishment shall be from a source approved, or considered satisfactory, by the health authority and which is in compliance with applicable State and local laws and regulations;

(b) food from such sources shall have been protected from contamination and spoilage during subsequent handling, packaging, and storage, and while in transit;

(c) all food in the food-service establishment shall be wholesome and free from spoilage, adulteration, and misbranding.

(2) Milk and Milk Products:

(a) all milk and milk products, including fluid milk, other fluid dairy products and manufactured milk products, shall meet the standards of quality established for such products by applicable State and local laws and regulations;

(b) only pasteurized fluid milk and fluid-milk products shall be used or served. Dry milk and milk products may be reconstituted in the establishment if used for cooking purposes only;

(c) all milk and fluid-milk products for drinking purposes shall be purchased and served in the original, individual container in which they were packaged at the milk plant, or shall be served from an approved bulk milk dispenser: Provided, that cream, whipped cream or half and half, which is to be consumed on the premises, may be served from the original container of not more than one-half gallon capacity or from a dispenser approved by the health authority for such service, and for mixed drinks requiring less than one-half pint of milk, milk may be poured from one-quart or one-half gallon containers packaged at a milk plant.

(3) Frozen Desserts:

(a) all frozen desserts such as ice cream, soft frozen desserts, ice milk, sherbets, ices, and mix shall meet the standards of quality established for such products by applicable State and local laws and regulations.

(4) Shellfish:

(a) all oysters, clams, and mussels shall be from sources approved by the State shellfish authority: Provided, that if the source is outside the State, it shall be one which is certified by the State of origin;

(b) shell stock shall be identified with an official tag giving the name and certificate number of the original shell stock shipper and the kind and quantity of shell stock;

(c) fresh and frozen shucked oysters, clams, and mussels, shall be packed in nonreturnable containers identified with the name and address of the packer, repacker, or distributor, and the certificate number of the packer or repacker preceded by the abbreviated name of the State;

(d) shucked shellfish shall be kept in the original container until used.

(5) Meat and Meat Products:

(a) all meat and meat products shall have been inspected for wholesomeness under an official regulatory program; provided, that the health authority may accept other sources which are in his opinion satisfactory and which are in compliance with applicable State and local laws and regulations.

(6) Poultry and Poultry Meat Products:

(a) all poultry and poultry meat products shall have been inspected for wholesomeness under an official regulatory program; Provided, that the health authority may accept other sources which are in his opinion satisfactory and which are in compliance with applicable State and local laws and regulations.

(7) Bakery Products:

(a) all bakery products shall have been prepared in the food-service establishment or in a food-processing establishment: Provided, that the health authority may accept other sources which are in his opinion satisfactory and which are in compliance with applicable State and local laws and regulations;

(b) all cream-filled and custard-filled pastries shall have been prepared and handled in accordance with the requirements of Rule 290-5-14-.05(6).

(8) Hermetically Sealed Food:

(a) all hermetically sealed food shall have been processed in food-processing establishments.

Authority Ga. L. 1964, pp. 507, 559. **Administrative History.** Original Rule entitled "Permits Required" was filed and effective on July 19, 1965 as 270-5-6-.03. **Amended:** Rule repealed and a new Rule entitled "Food Supplies" adopted. Filed January 24, 1967; effective February 12, 1967. **Amended:** Rule renumbered as 290-5-14-.03. Filed June 10, 1980; effective June 30, 1980.

290-5-14-.04 Food Protection. Amended.

(1) All food, while being stored, prepared, displayed, served, or sold in food-service establishments, or transported between such establishments, shall be protected against contamination from dust, flies, roaches, rodents, and other vermin; unclean utensils and work surfaces; unnecessary handling; coughs and sneezes; flooding, drainage, and overhead leakage; and any other source of contamination.

(2) Conveniently located refrigeration facilities, hot food storage and display facilities, and effective insulated facilities, shall be provided as needed to assure the maintenance of all food at required temperatures during storage, preparation, display, and service. Each cold-storage facility used for the storage of perishable food in non-frozen state shall be provided with an indicating thermometer accurate to $+2°F$., located in the warmest part of the facility in which food is stored, and of such type and so situated that the thermometer can be easily and readily observed for reading.

(3) Temperatures:

(a) all perishable food shall be stored at such temperatures as will protect against spoilage;

(b) all potentially hazardous food shall, except when being prepared and served, and when being displayed for service, be kept at 45°F. or below, or 140°F. or above;

(c) all potentially hazardous food, when placed on display for service, shall be kept hot or cold as required hereafter:

1. If served hot, the temperature of such food shall be kept at 140°F. or above;

2. If served cold, such food shall be:

(i) Displayed in or on a refrigerated facility which can reduce or maintain the product temperature at 45° or below; or

(ii) Prechilled to a temperature of 45°F. or below, when placed on display for service, and the food temperature shall at no time during the display period exceed 55°F.

(d) following preparation, hollandaise and other sauces may be exempt from the temperature requirements of subsection (3)(c) of 290-5-14-.04 if they are prepared from fresh ingredients and are discarded as waste within three hours after preparation. Where such sauces require eggs as an ingredient, only fresh shell eggs shall be used.

(e) frozen food shall be kept at such temperatures as to remain frozen, except when being thawed for preparation or use. Potentially hazardous frozen food shall be thawed at refrigerator temperatures of 45°F. or below; or under cool, potable running water (70°F. or below); or quick-thawed as part of the cooking process; or by any other method satisfactory to the health authority.

Authority Ga. L. 1964, pp. 507, 559. **Administrative History.** Original Rule entitled "Application; Filing of" was filed and effective on July 19, 1965 as 270-5-6-.04. **Amended:** Rule repealed and a new Rule entitled "Food Protection" adopted. Filed January 24, 1967; effective February 12, 1967. **Amended:** Rule renumbered as 290-5-14-.04. Filed June 10, 1980; effective June 30, 1980.

290-5-14-.05 Preparation. Amended.

(1) Convenient and suitable utensils, such as forks, knives, tongs, spoons, or scoops, shall be provided and used to minimize handling of food at all points where food is prepared.

(2) All raw fruits and vegetables shall be washed thoroughly before being cooked or served.

(3) Stuffings, poultry, and stuffed meats and poultry, shall be heated, throughout, to a minimum temperature of 165°F., with no interruption of the cooking process.

(4) Pork and pork products shall be thoroughly cooked.

(5) Meat salads, poultry salads, potato salad, egg salad, cream-filled pastries, and other potentially hazardous prepared food shall be prepared (preferably from chilled products) with a minimum of manual contact, and on surfaces and with utensils which are clean and which, prior to use, have been sanitized.

(6) Custards, cream fillings, or similar products which are prepared by hot or cold processes, and which are used as puddings or pastry fillings, shall be kept at safe temperatures, except during necessary periods of preparation and service, and shall meet the following requirements as applicable:

(a) pastry fillings shall be placed in shells, or other baked goods either while hot (not less than 140°F.) or immediately following preparation, if a cold process is used; or

(b) such fillings and puddings shall be refrigerated at 45°F. or below in shallow pans, immediately after cooking or preparation, and held thereat until combined into pastries, or served;

(c) all completed custard-filled and cream-filled pastries shall, unless served immediately following filling, be refrigerated at 45°F. or below promptly after preparation, and held thereat pending service.

Authority Ga. L. 1964, pp. 507, 559. **Administrative History.** Original Rule entitled "Permits, Granting and Validity" was filed and effective on July 19, 1965 as 270-5-6-.05. **Amended:** Rule repealed and a new Rule entitled "Preparation" adopted. Filed January 24, 1967; effective February 12, 1967. **Amended:** Rule renumbered as 290-5-14-.05. Filed June 10, 1980; effective June 30, 1980.

290-5-14-.06 Storage. Amended.

(1) Containers of food shall be stored above the floor, on clean racks, dollies, or other clean surfaces, in such a manner as to be protected from splash and other contamination.

(2) Food not subject to further washing or cooking before serving shall be stored in such a manner as to be protected against contamination from food requiring washing or cooking.

(3) Wet storage of packaged food shall be prohibited.
Authority Ga. L. 1964, pp. 507, 559. **Administrative History.** Original Rule entitled "Display of Permit" was filed and effective on July 19, 1965 as 270-5-6-.06. **Amended:** Rule repealed and a new Rule entitled "Storage" adopted. Filed January 24, 1967; effective February 12, 1967. **Amended:** Rule renumbered as 290-5-14-.06. Filed June 10, 1980; effective June 30, 1980.

290-5-14-.07 Display and Service. Amended.

(1) Where unwrapped food is placed on display in all types of food service operations, including smorgasbords, buffets, and cafeterias, it shall be protected against contamination from customers and other sources by effective, easily cleanable, counter-protector devices, cabinets, display cases, containers, or other similar type of protective equipment. Self-service openings in counter guards shall be so designed and arranged as to protect food from manual contact by customers.

(2) Tongs, forks, spoons, picks, spatulas, scoops, and other suitable utensils shall be provided and shall be used by employees to reduce manual contact with food to a minimum. For self-service by customers, similar implements shall be provided.

(3) Dispensing scoops, spoons, and dippers, used in serving frozen desserts, shall be stored, between uses, either in an approved running-water dipper well, or in a manner approved by the health authority.

(4) Sugar shall be provided only in closed dispensers or in individual packages.

(5) Individual portions of food once served to a customer shall not be served again; provided, that wrapped food, other than potentially hazardous food, which is still wholesome and has not been unwrapped may be re-served.
Authority Ga. L. 1964, pp. 507, 559. **Administrative History.** Original Rule entitled "Processing of Application" was filed and effective on July 19, 1965 as 270-5-6-.07. **Amended:** Rule repealed and a new Rule entitled "Display and Service" adopted. Filed January 24, 1967; effective February 12, 1967. **Amended:** Rule renumbered as 290-5-14-.07. Filed June 10, 1980; effective June 30, 1980.

290-5-14-.08 Transportation. Amended.

(1) The requirements for storage, display, and general protection against contamination, as contained in this subsection, shall apply in the transporting of all food from a food-service establishment to another location for service or catering operations and all potentially hazardous food shall be kept at 45°F. or below, or 140°F. or above, during transportation.

(2) During the transportation of food from a food-service establishment, all food shall be in covered containers or completely wrapped or packaged so as to be protected from contamination.
Authority Ga. L. 1964, pp. 507, 559. **Administrative History.** Original Rule entitled "Inspection of Premises" was filed and effective on July 19, 1965 as 270-5-6-.08. **Amended:** Rule repealed and a new Rule entitled "Transportation" adopted. Filed January 24, 1967; effective February 12, 1967. **Amended:** Rule renumbered as 290-5-14-.08. Filed June 10, 1980; effective June 30, 1980.

290-5-14-.09 Poisonous and Toxic Materials. Amended.

(1) Only those poisonous and toxic materials required to maintain the establishment in a sanitary condition, and for sanitization of equipment and utensils, shall be present in any area used in connection with food-service establishments.

(2) All containers of poisonous and toxic materials shall be prominently and distinctively marked or labeled for easy identification as to contents.

(3) Poisonous and toxic materials shall be stored in cabinets which are used for no other purpose, or in a place which is outside the food-storage, food-preparation, and cleaned equipment and utensil storage rooms. Bactericides and cleaning compounds shall not be stored in the same cabinet or area of the room with insecticides, rodenticides, or other poisonous materials.

(4) Bactericides, cleaning compounds, or other compounds, intended for use on food-contact surfaces, shall not be used in such a manner as to leave a toxic residue on such surfaces, nor to constitute a hazard to employees or customers.

(5) Poisonous polishing materials shall not be used on equipment or utensils, nor stored in the establishment.

(6) Poisonous compounds, such as insecticides and rodenticides, in powdered form, shall have a distinctive color so as not to be mistaken for food.

(7) Poisonous materials shall not be used in any way as to contaminate food, equipment, or utensils, nor to constitute other hazards to employees or customers.
Authority Ga. L. 1964, pp. 507, 559. **Administrative History.** Original Rule entitled "Routine Inspection" was filed and effective on July 19, 1965 as 270-5-6-.09. **Amended:** Rule repealed and a new Rule entitled "Poisonous and Toxic Materials" adopted. Filed January 24, 1967; effective February 12, 1967. **Amended:** Rule renumbered as 290-5-14-.09. Filed June 10, 1980; effective June 30, 1980.

290-5-14-.10 Personnel. Amended.

(1) Health and Disease Control:

(a) disease control

1. No person while affected with a disease in a communicable form, or while a carrier of such disease, or while afflicted with boils, infected wounds, or an acute infection, shall work in a food-service establishment in an area and capacity in which there is a likelihood of transmission of disease to patrons or to fellow employees, either through direct contact or through the contamination of food or food-contact surfaces with pathogenic organisms. No person shall be employed in such an area and capacity in a food-service establishment.

(b) reporting

1. The manager or person in charge of the establishment shall notify the health authority when any employee of a food-service establishment is known or suspected of having a disease in a communicable form.

(c) cleanliness

1. Employees shall maintain a high degree of personal cleanliness and shall conform to good hygienic practices during all working periods.

2. Handwashing

(i) All employees shall thoroughly wash their hands and arms with soap and warm water before starting work, and shall wash hands during work hours as often as may be required to remove soil and contamination, as well as after visiting the toilet room.

RULES OF GEORGIA DEPARTMENT OF AGRICULTURE FOOD DIVISION REGULATIONS

Chapter 40-7-5: Additional Regulations Applicable to Processing Plants

Table of Contents

40-7-5-.01 Scope of Regulations. These regulations shall apply to all plants where food is processed, which are not covered elsewhere in these regulations.
Authority Ga. L. 1956, p. 195 as amended. **Administrative History.** Original Rule filed June 30, 1965.

40-7-5-.02 Floors. Where water comes in contact with the floor during processing and clean-up operations, the floors shall be sloped toward an adequate drain system.
Authority Ga. L. 1956, p. 195 as amended. **Administrative History.** Original Rule filed June 30, 1965.

40-7-5-.03 Water. When water is re-used during the processing of food products, it shall be re-used backwards in the flow plan under strict sanitary conditions.
Authority Ga. L. 1956, p. 195 as amended. **Administrative History.** Original Rule filed June 30, 1965.

40-7-5-.04 Clean Utensils and Equipment.

(1) All utensils and equipment shall be thoroughly cleaned and sanitized as necessary.

(2) Sinks and all automatic washing equipment shall be sufficiently large enough so that all utensils can be easily cleaned and sanitized as necessary.

(3) Processing plants must have an adequate water supply with hot and cold running water, under pressure, with a minimum hot water temperature of 140 degrees F., available at all times wherever needed.
Authority Ga. L. 1956, p. 195 as amended. **Administrative History.** Original Rule filed June 30, 1965.

40-7-5-.05 Refrigeration.

(1) All readily perishable foods requiring refrigeration shall be kept at an air temperature of 40 degrees Fahrenheit or below.

(2) All frozen foods shall be kept at an air temperature of zero degrees F. or less during storage, and shall not exceed a product temperature of 10 degrees F. increase for a reasonable time during transit from processing plant to other storage areas or point of sale.

(3) Those frozen foods to be used for further processing and/or to be repackaged in smaller units, may be defrosted for such purposes in accordance with good processing and sanitary practices.
Authority Ga. L. 1956, p. 195 as amended. **Administrative History.** Original Rule filed June 30, 1965.

TEXAS DEPARTMENT OF HEALTH. DIVISION OF FOOD AND DRUGS. RULES ON FOOD SERVICE SANITATION.

(b) Food Protection.

(1) General. At all times, including while being stored, prepared, displayed, served, or transported, food shall be protected from potential contamination, including dust, insects, rodents, unclean equipment and utensils, unnecessary handling, coughs and sneezes, flooding, drainage, and overhead leakage or overhead drippage from condensation. The internal temperature of potentially hazardous food shall be 45°F. (7°C.) or below or 140°F. (60°C.) or above at all times, except as otherwise provided in these rules.

(2) Emergency Occurrences. In the event of an occurrence, such as a fire, flood, power outage, or similar event, which might result in the contamination of food, or which might prevent potentially hazardous food from being held at required temperatures, the person in charge shall immediately contact the regulatory authority. Upon receiving notice of this occurrence, the regulatory authority shall take whatever action that it deems necessary to protect the public health.

(c) Food Storage.

(1) General.

a) Food, whether raw or prepared, if removed from the container or package in which it was obtained, shall be stored in a clean covered container except during necessary periods of preparation or service. Container covers shall be impervious and nonabsorbent, except that linens or napkins may be used for lining or covering bread or roll containers. Solid cuts of meat shall be protected by being covered in storage, except that quarters or sides of meat may be hung uncovered on clean sanitized hooks if no food product is stored beneath the meat.

b) Containers of food shall be stored a minimum of six inches above the floor in the manner that protects the food from splash and other contamination, and that permits easy cleaning of the storage area, except that:

(i) Metal pressurized beverage containers, and cased food packaged in cans, glass or other waterproof containers need not be elevated when the food container is not exposed to floor moisture; and

(ii) Containers may be stored on dollies, racks or pallets, provided such equipment is easily movable.

c) Food and containers of food shall not be stored under exposed unprotected sewer lines or waterlines, except for automatic fire protection sprinkler heads that may be required by law. The storage of food in toilet rooms or vestibules is prohibited.

d) Food not subject to further washing or cooking before serving shall be stored in a way that protects it against cross-contamination from food requiring washing or cooking.

e) Packaged food shall not be stored in contact with water or undrained ice. Wrapped sandwiches shall not be stored in direct contact with ice.

f) Unless its identity is unmistakable, bulk food such as cooking oil, syrup, salt, sugar or flour not stored in the product container or package in which it was obtained, shall be stored in a container identifying the food by common name.

(2) Refrigerated Storage.

a) Conveniently located refrigeration facilities or effectively insulated facilities shall be provided to assure the maintenance of all potentially hazardous food at required temperatures during storage. Each mechanically refrigerated facility storing potentially hazardous food shall be provided with a numerically scaled indicating thermometer, accurate to $\pm 3°F$. ($1.6°C$.), located to measure the air temperature in the warmest part of the facility and located to be easily readable. Recording thermometers, accurate to $\pm 3°F$. ($1.6°C$.), may be used in lieu of indicating thermometers.

b) Potentially hazardous food requiring refrigeration after preparation shall be rapidly cooled to an internal temperature of $45°F$. ($7°C$.) or below. Potentially hazardous foods of large volume or prepared in large quantities shall be rapidly cooled utilizing such methods as shallow pans, agitation, quick chilling or water circulation external to the food container so that the cooling period shall not exceed four (4) hours. Potentially hazardous food to be transported shall be prechilled and held at a temperature of $45°F$. ($7°C$.) or below unless maintained in accordance with paragraph (3 b) of subsection (c) of this rule.

c) Frozen foods shall be kept frozen and shall be stored at a temperature of $0°F$. ($-18°C$.) or below.

d) Ice intended for human consumption shall not be used as a medium for cooling stored food, food containers or food utensils, except that such ice may be used for cooling tubes conveying beverages or beverage ingredients to a dispenser head. Ice used for cooling stored food and food containers shall not be used for human consumption.

(3) Hot Storage.

a) Conveniently located hot food storage facilities shall be provided to assure the maintenance of food at the required temperature during storage. Each hot food facility storing potentially hazardous food shall be provided with a numerically scaled indicating thermometer, accurate to $\pm 3°F$. ($1.6°C$.), located to measure the air temperature in the coolest part of the facility and located to be easily readable. Recording thermometers, accurate to $\pm 3°F$. ($1.6°C$.), may be used in lieu of indicating thermometers. Where it is impractical to install thermometers on equipment such as bainmaries, steam tables, steam kettles, heat lamps, cal-rod units, or insulated food transport carriers, a product thermometer must be available and used to check internal food temperature.

b) The internal temperature of potentially hazardous foods requiring hot storage shall be 140°F. (60°C.) or above except during necessary periods of preparation. Potentially hazardous food to be transported shall be held at a temperature of 140°F. (60°C.) or above unless maintained in accordance with paragraph (2 b) of subsection (c) of this rule.

(d) Food Preparation.

(1) General. Food shall be prepared with the least possible manual contact, with suitable utensils, and on surfaces that prior to use have been cleaned, rinsed and sanitized to prevent cross-contamination.

(2) Raw Fruits and Raw Vegetables. Raw fruits and raw vegetables shall be thoroughly washed with potable water before being cooked or served.

(3) Cooking Potentially Hazardous Foods. Potentially hazardous foods requiring cooking shall be cooked to heat all parts of the food to a temperature of at least 140°F. (60°C.), except that:

a) Poultry, poultry stuffings, stuffed meats and stuffings containing meat shall be cooked to heat all parts of the food to at least 165°F. (74°C.) with no interruption of the cooking process.

b) Pork and any food containing pork shall be cooked to heat all parts of the food to at least 150°F. (66°C.).

c) Rare roast beef shall be cooked to an internal temperature of at least 130°F. (54°C.) and rare beef steak shall be cooked to a temperature of 130°F. (54°C.) unless otherwise ordered by the immediate consumer.

(4) Dry Milk and Dry Milk Products. Reconstituted dry milk and dry milk products may be used in instant desserts and whipped products, or for cooking and baking purposes.

(5) Liquid, Frozen, Dry Eggs and Egg Products. Liquid, frozen, dry eggs and egg products shall be used only for cooking and baking purposes.

(6) Reheating. Potentially hazardous foods that have been cooked and then refrigerated, shall be reheated rapidly to 165°F. (74°C.) or higher throughout before being served or before being placed in a hot food storage facility. Steam tables, bainmaries, warmers, and similar hot food holding facilities are prohibited for the rapid reheating of potentially hazardous foods.

(7) Nondairy Products. Nondairy creaming, whitening, or whipping agents may be reconstituted on the premises only when they will be stored in sanitized, covered containers not exceeding one gallon in capacity and cooled to 45°F. (7°C.) or below within four hours after preparation.

(8) Product Thermometers. Metal stem-type numerically scaled indicating thermometers, accurate to ±2°F. (1.1°C.) shall be provided and used to assure the attainment and maintenance of proper internal cooking, holding, or refrigeration temperatures of all potentially hazardous foods.

(9) Thawing Potentially Hazardous Foods. Potentially hazardous foods shall be thawed:

a) In refrigerated units at a temperature not to exceed 45°F. (7°C.); or

b) Under potable running water of a temperature of 70°F. (21°C.) or below, with sufficient water velocity to agitate and float off loose food particles into the overflow; or

c) In a microwave oven only when the food will be immediately transferred to conventional cooking facilities as part of a continuous cooking process or when the entire, uninterrupted cooking process takes place in the microwave oven; or

d) As part of the conventional cooking process.

(e) Food Display and Service.

(1) Potentially Hazardous Foods. Potentially hazardous food shall be kept at an internal temperature of 45°F. (7°C.) or below or at an internal temperature of 140°F. (60°C.) or above during display and service, except that rare roast beef shall be held for service at a temperature of at least 130°F. (54°C.).

(2) Milk and Cream Dispensing.

a) Milk and milk products for drinking purposes shall be provided to the consumer in an unopened, commercially filled package not exceeding one pint in capacity, or drawn from a commercially filled container stored in a mechanically refrigerated bulk milk dispenser. Where it is necessary to provide individual servings under special institutional circumstances, milk and milk products may be poured from a commercially filled container provided such a procedure is authorized by the regulatory authority. Where a bulk dispenser for milk and milk products is not available and portions of less than one-half pint are required for mixed drinks, cereal, or dessert service, milk and milk products may be poured from a commercially filled container.

b) Cream or half and half shall be provided in an individual service container, protected pour-type pitcher, or drawn from a refrigerated dispenser designed for such service.

(3) Nondairy Product Dispensing. Nondairy creaming or whitening agents shall be provided in an individual service container, protected pour-type pitcher, or drawn from a refrigerated dispenser designed for such service.

(4) Condiment Dispensing.

a) Condiments, seasonings and dressings for self-service use shall be provided in individual packages, from dispensers, or from containers protected in accordance with paragraph (8) of subsection (e) of this rule.

b) Condiments provided for table or counter service shall be individually portioned, except that catsup and other sauces may be served in the original container or pour-type dispenser. Sugar for consumer usage shall be provided in individual packages or in pour-type dispensers.

(5) Ice Dispensing. Ice for consumer use shall be dispensed only by employees with scoops, tongs, or other ice-self-dispensing utensils or through automatic service, ice-dispensing equipment. Ice dispensing utensils shall be stored on a clean surface or in the ice with the dispensing utensil's handle extended out of the ice. Between uses, ice transfer receptacles shall be stored in a way that protects them from contamination. Ice storage bins shall be drained through an air gap.

(6) Dispensing Utensils. To avoid unnecessary manual contact with food, suitable dispensing utensils shall be used by employees or provided to consumers who serve themselves. Between uses during service, dispensing utensils shall be:

a) Stored in the food with the dispensing utensil handle extended out of the food; or

b) Stored clean and dry; or

c) Stored in running water; or

d) Stored either in a running water dipper well, or clean and dry in the case of dispensing utensils and malt collars used in preparing frozen desserts.

(7) Reservice. Once served to a consumer, portions of leftover food shall not be served again except that packaged food, other than potentially hazardous food, that is still packaged and is still in sound condition, may be re-served.

(8) Display Equipment. Food on display shall be protected from consumer contamination by the use of packaging or by the use of easily cleanable counter, serving line or salad bar protector devices, display cases, or by other effective means. Enough hot or cold food facilities shall be available to maintain the required temperature of potentially hazardous food on display.

(9) Reuse of Tableware. Reuse of soiled tableware by self-service consumers returning to the service area for additional food is prohibited.

(10) Food Transportation. During transportation, food and food utensils shall be kept in covered containers or completely wrapped or packaged so as to be protected from contamination. Foods in original individual packages do not need to be overwrapped or covered if the original package has not been torn or broken. During transportation, including transportation to another location for service or catering operations, food shall meet the requirements of these rules relating to food protection and food storage.

Appendix IV

Representative Foodservice Establishment Inspection Report

TEXAS DEPARTMENT OF HEALTH
Division of Food and Drug

Est. I.D. [1-10]	County	Dist.	Est.No.	Census Tract 11-13	Sanit.Code 14-16	17-22 Yr.	Mo.	Day	Travel Time 23-25	Inspec.Time 26-28

Owner Name: _____ Establishment Name: _____

Address: _____ Zip _____

PURPOSE	Food Service Establishment Inspection Report
Regular .... 29-1 Follow-up .... 2 Complaint .... 3 Investigation ..... 4 Other 5	Based on an inspection this day, the items circled below identify the violations in operations or facilities which must be corrected by the next routine inspection or such shorter period of time as may be specified in writing by the regulatory authority. Failure to comply with any time limits for corrections specified in this notice may result in cessation of your Food Service operations.

FOOD

ITEM NO.		WT.	COL.
*01	Source; sound condition, no spoilage	5	30
02	Original container; properly labeled	1	31

FOOD PROTECTION

		WT.	COL.
*03	Potentially hazardous food meets temperature requirements during storage, preparation, display, service transportation	5	32
*04	Facilities to maintain product temperature	4	33
05	Thermometers provided and conspicuous	1	34
06	Potentially hazardous food properly thawed	2	35
*07	Unwrapped and potentially hazardous food not re-served	4	36
08	Food protection during storage, preparation, display, service, transportation	2	37
09	Handling of food (ice) minimized	2	38
10	In use, food (ice) dispensing utensils properly stored	1	39

PERSONNEL

		WT.	COL.
*11	Personnel with infections restricted	5	40
*12	Hands washed and clean, good hygienic practices	5	41
13	Clean clothes, hair restraints	1	42

FOOD EQUIPMENT & UTENSILS

		WT.	COL.
14	Food (ice) contact surfaces: designed, constructed, maintained, installed, located	2	43
15	Non-food contact surfaces: designed, constructed, maintained, installed, located	1	44
16	Dishwashing facilities: designed, constructed, maintained, installed, located, operated	2	45
17	Accurate thermometers, chemical test kits provided, gauge cock (¼" IPS valve)	1	46
18	Pre-flushed, scraped, soaked	1	47
19	Wash, rinse water: clean, proper temperature	2	48
*20	Sanitization rinse: clean, temperature, concentration, exposure time; equipment, utensils sanitized	4	49
21	Wiping cloths: clean, use/restricted	1	50
22	Food-contact surfaces of equipment and utensils clean, free of abrasives, detergents	2	51
23	Non-food contact surfaces of equipment and utensils clean	1	52
24	Storage, handling of clean equipment/utensils	1	53
25	Single-service articles, storage, dispensing	1	54
26	No re-use of single service articles	2	55

WATER

		WT.	COL.
*27	Water source, safe: hot & cold under pressure	5	56

SEWAGE

		WT.	COL.
*28	Sewage and waste water disposal	4	57

PLUMBING

		WT.	COL.
29	Installed, maintained	1	58
*30	Cross-connection, back siphonage, backflow	5	59

TOILET & HANDWASHING FACILITIES

		WT.	COL.
*31	Number, convenient, accessible, designed, installed	4	60
32	Toilet rooms enclosed, self-closing doors, fixtures, good repair, clean; hand cleanser, sanitary towels/tissues/hand-drying devices provided, proper waste receptacles	2	61

GARBAGE & REFUSE DISPOSAL

		WT.	COL.
33	Containers or receptacles, covered: adequate number insect/rodent proof, frequency, clean	2	62
34	Outside storage area enclosures properly constructed, clean, controlled incineration	1	63

INSECT, RODENT, ANIMAL CONTROL

		WT.	COL.
*35	Presence of insects/rodents — outer openings protected, no birds, turtles, other animals	4	64

FLOORS, WALLS & CEILINGS

		WT.	COL.
36	Floors, constructed, drained, clean, good repair, covering installation, dustless cleaning methods	1	65
37	Walls, ceiling, attached equipment: constructed, good repair, clean, surfaces, dustless cleaning methods	1	66

LIGHTING

		WT.	COL.
38	Lighting provided as required, fixtures shielded	1	67

VENTILATION

		WT.	COL.
39	Rooms and equipment — vented as required	1	68

DRESSING ROOMS

		WT.	COL.
40	Rooms clean, lockers provided, facilities clean, located	1	69

OTHER OPERATIONS

		WT.	COL.
*41	Toxic items properly stored, labeled, used	5	70
42	Premises maintained free of litter, unnecessary articles, cleaning maintenance equipment properly stored. Authorized personnel	1	71
43	Complete separation from living/sleeping quarters. Laundry.	1	72
44	Clean, soiled linen properly stored	1	73

FOLLOW-UP	RATING SCORE 75-77	Type of Est.
Yes .... 74-1 No 2	100 less weight of items violated ➤	78-80

*Critical Items Requiring Immediate Attention.   Remarks on back [80-1]

Received by: name _____

title _____

Inspected by: name _____

title _____

FORM FV-416-1
(6-1-61)

U.S. DEPARTMENT OF AGRICULTURE
CONSUMER AND MARKETING SERVICE

SANITATION SCORE SHEET FOR CANNED FOOD PROCESSING PLANTS

PLANT	LOCATION	DATE	D.I.R. NO.

RATING SYMBOLS:
S – Satisfactory
NI – Needs improvement
U – Unsatisfactory

TIME OF INSPECTION ▶

EXPLANATION

AREA AND ITEMS	S	NI	U	S	NI	U	S	NI	U	EXPLANATION
PREMISES										
1 Outside areas										
2 Waste disposal										
3										
RECEIVING DEPARTMENT										
1 Boxes										
2 Storage										
3 Dumpers & conveyors										
4 Floors, gutters & walls										
5										
PREPARATION DEPARTMENT										
1 Washers & flumes										
2 Belts & elevators										
3 Graders										
4 Cutters & slicers										
5 Blanchers										
6 Pulpers & finishers										
7 Floors, gutters & walls										
8										
CANNING DEPARTMENT										
1 Belts										
2 Fillers & can tables										
3 Floors, gutters & walls										
4										
COOK ROOM										
1 Exhaust box										
2 Syrupers										
3 Seamers										
4 Floors, gutters & walls										
5										
SYRUP AND EVAPORATION DEPARTMENT										
1 Tanks & pipes										
2 Vacuum pans										
3 Floors, gutters & walls										
4										
WAREHOUSE										
1 General housekeeping										
2 Stacks										
3										
REST ROOMS										
1 Supplies										
2 Wash basins										
3 Toilets & urinals										
4 Floors & walls										
5										
PERSONNEL										
1 Cleanliness										
2 Head covering										
3 Smoking										
4										

See reverse for additional remarks.

OVERALL SANITATION RATING

(USDA INSPECTOR)

EXHIBIT 1

FORM FV-416-1
(5-76)

U.S. DEPARTMENT OF AGRICULTURE

SANITATION SCORE SHEET FOR CANNED FOOD PROCESSING PLANTS

NAME OF PLANT	LOCATION	DATE	D.I.R. NO.
ABC	XYZ	7/8/79	10

RATING SYMBOLS (√) Satisfactory
MN – Minor MJ – Major
CR – Critical U – Unsatisfactory

SIGNATURE OF INSPECTOR(S)

J. Beam

SANITATION DEFICIENCIES - SHOW ITEM, NO. AND DESCRIBE

	TIME	0600	1700	

PREMISES

		0600	1700	DEFICIENCIES
A	1. Outside areas	x	x	A-2 Dirty trash barrels on dock area (corrected 0700)
	2. Waste disposal	MN	x	
	3.			

RECEIVING DEPARTMENT

		0600	1700	
B	1. Boxes	x	x	B-4 Excessive product spillage – broken bins (corrected immediately).
	2. Storage	x	x	
	3. Dumpers & conveyors	x	x	
	4. Floors, gutters & walls	MN	x	
	5.			

PREPARATION DEPARTMENT

		0600	1700	
C	1. Washers & flumes	MJ	OK	C-1 C-1 Slime on side of flume #6.
	2. Belts & Elevators	MJ	x	
	3. Graders & Snippers	x	x	C-2 Slime on belt after scrubber to sorting (corrected 0730).
	4. Cutters & slicers	x	x	
	5. Blanchers, hoppers & de-waterers	x	x	
	6. Pulpers & finishers	x	x	
	7. Floors, gutters & walls	x	x	
	8.			
	9.			
	10.			

CANNING DEPARTMENT

		0600	1700	
D	1. Belts	MJ	U	D-1 Slime on canning belt #6 – informed plant management.
	2. Fillers & can tables	x	x	D-1 No corrective action taken between shifts. Unsatisfactory sanitation.
	3. Floors, gutters & walls	MN	OK	D-3 Excessive product on floor by line #4.
	4.			
	5.			

COOK ROOM

		0600	1700
E	1. Exhaust box	x	x
	2. Syrupers	x	x
	3. Seamers	x	x
	4. Floors, gutters & walls	x	x
	5.		

SYRUP & EVAPORATION DEPT.

		0600	1700
F	1. Tanks & pipes	x	x
	2. Vacuum pans	x	x
	3. Floors, gutters & walls	x	x
	4.		

WAREHOUSE

		0600	1700
G	1. General housekeeping	x	x
	2. Stacks	x	x
	3.		

REST ROOMS

		0600	1700
H	1. Supplies	x	x
	2. Wash basins	x	x
	3. Toilets & urinals	x	x
	4. Floors & walls	x	x
	5.		

PERSONNEL

		0600	1700	
I	1. Cleanliness	x		Personnel hygiene program verified. All areas in compliance.
	2. Head covering	x		
	3. Smoking	x		
	4.			

Appendix V

FDA-Accepted Sanitizing Agents and Use Regulations

(Taken from Code of Federal Regulations, 1977)

§178.1010 SANITIZING SOLUTIONS

Sanitizing solutions may be safely used on food-processing equipment and utensils, and on other food-contact articles as specified in this section, within the following prescribed conditions:

(a) Such sanitizing solutions are used, followed by adequate draining, before contact with food.

(b) The solutions consist of one of the following, to which may be added components generally recognized as safe and components which are permitted by prior sanction or approval.

(1) An aqueous solution containing potassium, sodium, or calcium hypochlorite, with or without the bromides of potassium, sodium, or calcium.

(2) An aqueous solution containing dichloroisocyanuric acid, trichloroisocyanuric acid, or the sodium or potassium salts of these acids, with or without the bromides of potassium, sodium, or calcium.

(3) An aqueous solution containing potassium iodide, sodium p-toluenesulfonchloroamide, and sodium lauryl sulfate.

(4) An aqueous solution containing iodine, butoxy monoether of mixed (ethylene-propylene) polyalkylene glycol having a cloudpoint of 90°—100°C in 0.5 percent aqueous solution and an average molecular weight of 3,300, and ethylene glycol monobutyl ether. Additionally, the aqueous solution may contain diethylene glycol monoethyl ether as an optional ingredient.

(5) An aqueous solution containing elemental iodine, hydriodic acid, α-(p-nenylphenyl)-omega-hydroxypoly(oxyethylene) (complying with the identity described in 178.3400(c) and having a maximum average molecular weight of 748) and/or polyoxyethylene-polyoxypropylene block polymers (having a minimum average molecular weight of 1,900). Additionally, the aqueous solution may contain isopropyl alcohol as an optional ingredient.

(6) An aqueous solution containing elemental iodine, sodium iodide, sodium dioctylsulfosuccinate, and polyoxyethylene-polyoxypropylene block polymers (having a minimum average molecular weight of 1,900).

(7) An aqueous solution containing dodecylbenzenesulfonic acid, polyoxyethylene-polyoxypropylene block polymers (having a minimum average molecular weight of 2,800). In addition to use on food-processing equipment and utensils, this solution may be used on glass bottles and other glass containers intended for holding milk.

(8) An aqueous solution containing elemental iodine, butoxy mono-ether of mixed (ethylene-propylene) polyalkylene glycol having a minimum average molecular weight of 2,400 and α-lauroyl-ω-hydroxypoly(oxyethylene) with an average 8–9 moles of ethylene oxide and an average molecular weight of 400. In addition to use on food-processing equipment and utensils, this solution may be used on beverage containers, including milk containers or equipment. Rinse water treated with this solution can be recirculated as a preliminary rinse. It is not to be used as final rinse.

(9) An aqueous solution containing n-alkyl (C_{12}—C_{15}) benzyldimethylammonium chloride compounds having average molecular weights of 351–380 and consisting principally of alkyl groups with 12–16 carbon atoms with or without not over 1 percent each of groups with 8 and 10 carbon atoms. Additionally, the aqueous solution may contain isopropyl alcohol as an optional ingredient.

(10) An aqueous solution containing trichloromelamine and either sodium lauryl sulfate or dodecylbenzenesulfonic acid. In addition to use on food-processing equipment and utensils and other food-contact articles, this solution may be used on beverage containers except milk containers or equipment.

(11) An aqueous solution containing equal amounts of n-alkyl (C_{12}—C_{15}) benzyl dimethyl ammonium chloride and n-alkyl (C_{12}—C_{15}) dimethyl ethylbenzyl ammonium chloride (having an average molecular weight of 384). In addition to use on food-processing equipment and utensils, this solution may be used on food-contact surfaces in public eating places.

(12) An aqueous solution containing the sodium salt of sulfonated oleic acid, polyoxyethylene-polyoxypropylene block polymers (having

an average molecular weight of 2,000 and 27 to 31 moles of polyoxy-propylene). In addition to use on food-processing equipment and utensils, this solution may be used on glass bottles and other glass containers intended for holding milk. All equipment, utensils, glass bottles, and other glass containers treated with this sanitizing solution shall have a drainage period of 15 minutes prior to use in contact with food.

(13) An aqueous solution containing elemental iodine and alkyl (C_{12}—C_{15}) monoether of mixed (ethylene-propylene) polyalkylene glycol, having a cloudpoint of 70°–77°C in 1 percent aqueous solution and an average molecular weight of 807.

(14) An aqueous solution containing iodine, butoxy monoether of mixed (ethylene-propylene) polyalkylene glycol, having a cloudpoint of 90°–100°C in 0.5 percent aqueous solution and an average molecular weight of 3,300, and polyoxyethylene-polyoxypropylene block polymers (having a minimum average molecular weight of 2,000).

(15) An aqueous solution containing lithium hypochlorite.

(16) An aqueous solution containing equal amounts of n-alkyl (C_{12}—C_{15}) benzyl dimethyl ammonium chloride and n-alkyl (C_{12}—C_{15}) dimethyl ethylbenzyl ammonium chloride (having average molecular weights of 377–384), with the optional adjuvant substances tetrasodium ethylene-diaminetraacetate and/or α-(p-nonylphenol)-ω-hydroxypoly-(oxyethylene) having an average poly(oxyethylene) content of 11 moles. In addition to use of food-processing equipment and utensils, this solution may be used on food-contact surfaces in public eating places.

(17) An aqueous solution containing di-n-alkyl (C_8—C_{10}) dimethyl ammonium chlorides and isopropyl alcohol, having average molecular weights of 332–361. In addition to use on food-processing equipment and utensils, this solution may be used on food-contact surfaces in public eating places.

(18) An aqueous solution containing n-alkyl (C_{12}—C_{18}) benzyldimethylammonium chloride, sodium metaborate, α-terpineol and α[p-(1,1,3,3-tetra-methylbutyl) phenyl]-ω-hydroxypoly(oxyethylene) produced with one mole of the phenol and 4 to 14 moles ethylene oxide.

(19) An aqueous solution containing sodium dichloroisocyanurate and tetra-sodium ethylenediaminetetraacetate. In addition to use on food-processing equipment and utensils, this solution may be used on food-contact surfaces in public eating places.

(20) An aqueous solution containing ortho-phenylphenol, ortho-benzyl-para-chlorophenol, para-tertiaryamylphenol, sodium-α-alkyl (C_{12}—C_{13})-ω-hydroxypoly(oxyethylene) sulfate with the poly(oxyethylene) content averaging one mole, potassium salts of coconut oil fatty acids, and isopropyl alcohol or hexylene glycol.

(21) An aqueous solution containing sodium dodecylbenzenesulfonate.

In addition to use on food-processing equipment and utensils, this solution may be used on glass bottles and other glass containers intended for holding milk.

(c) The solutions identified in paragraph (b) of this section will not exceed the following concentrations:

(1) Solutions identified in paragraph (b) (1) of this section will provide not more than 200 parts per million of available halogen determined as available chlorine.

(2) Solutions identified in paragraph (b) (2) of this section will provide not more than 100 parts per million of available halogen determined as available chlorine.

(3) Solutions identified in paragraph (b) (3) of this section will provide not more than 25 parts per million of titratable iodine. The solutions will contain the components potassium iodide, sodium p-toluenesulfonchloramide, and sodium lauryl sulfate at a level not in excess of the minimum required to produce their intended functional effect.

(4) Solutions identified in paragraph (b) (4), (5), (6), (8), (13), and (14) of this section will contain iodine to provide not more than 25 parts per million of titratable iodine. The adjuvants used with the iodine will not be in excess of the minimum amounts required to accomplish the intended technical effect.

(5) Solutions identified in paragraph (b) (7) of this section will provide not more than 400 parts per million of dodecylbenzenesulfonic acid and not more than 80 parts per million of polyoxyethylene-polyoxypropylene block polymers (having a minimum average molecular weight of 2,800).

(6) Solutions identified in paragraph (b) (9) of this section shall provide when ready to use no more than 200 parts per million of the active quaternary compound.

(7) Solutions identified in paragraph (b) (10) of this section shall provide not more than sufficient trichloromelamine to produce 200 parts per million of available chlorine and either sodium lauryl sulfate at a level not in excess of the minimum required to produce its intended functional effect or not more than 400 parts per million of dodecylbenzenesulfonic acid.

(8) Solutions identified in paragraph (b) (11) of this section shall provide, when ready to use, no more than 200 parts per million of active quaternary compound.

(9) The solution identified in paragraph (b) (12) of this section shall provide not more than 200 parts per million of sulfonated oleic acid, sodium salt.

(10) Solutions identified in paragraph (b) (15) of this section shall provide not more than 200 parts per million of available chlorine and not more than 30 ppm lithium.

(11) Solutions identified in paragraph (b) (16) of this section shall provide not more than 200 parts per million of active quaternary compound.

(12) Solutions identified in paragraph (b) (17) of this section shall provide, when ready to use, a level of 150 parts per million of the active quaternary compound.

(13) Solutions identified in paragraph (b) (18) of this section shall provide not more than 200 parts per million of active quaternary compound and not more than 66 parts per million of α [p-(1,1,3,3-tetramethylbutyl)phenyl]-ω-hydroxypoly(oxyethylene).

(14) Solutions identified in paragraph (b) (19) of this section shall provide, when ready to use, a level of 100 parts per million of available chlorine.

(15) Solutions identified in paragraph (b) (20) of this section are for single use applications only and shall provide, when ready to use, a level of 800 parts per million of total active phenols consisting of 400 parts per million *ortho*-phenyl-phenol, 320 parts per million *ortho*-benzyl-*para*-chlorophenol and 80 parts per million *para*-tertiaryamyl-phenol.

(16) Solution identified in paragraph (b) (21) of this section shall provide not more than 430 parts per million and not less than 25 parts per million of sodium dodecylbenzenesulfonate.

(d) Sanitizing agents for use in accordance with this section will bear labeling meeting the requirements of the Federal Insecticide, Fungicide, and Rodenticide Act.

Index